The Nicholson Mine

Saskatchewan's First Cold War Uranium Mine

LAURIER L. SCHRAMM

And

PATTY OGILVIE-EVANS

The Authors:
Dr. Laurier L. Schramm and Patty Ogilvie-Evans
Saskatchewan Research Council
125 – 15 Innovation Blvd.
Saskatoon, Canada, S7N 2X8

Print ISBN: 978-0-9958081-4-0
ePub ISBN: 978-0-9958081-5-7

SCHRAMM **and** OGILVIE-EVANS

DEDICATION

All proceeds from the sales of this book will go to the Saskatchewan Research Council's Technology in Action Fund - a perpetual memorial fund established to help the people of Saskatchewan develop their province as a highly skilled, fair, desirable and compassionate society with a secure environment through research, development and the transfer of innovative scientific and technological solutions, applications and services.

CONTENTS

PREFACE

The first discovery of uranium in Saskatchewan was at Nicholson Bay, in a remote location in northern Saskatchewan, on the shore of Lake Athabasca, the 22nd largest lake in the world. Uranium was first noted at what became the Nicholson site in 1929 when uranium was only of interest as an indicator of radium potential. When uranium ores became of strategic national interest in about 1940, a cross-Canada search was launched to find uranium deposits. The first to be found and developed was in the Northwest Territories. The second arose from a return to exploration at the Nicholson site in the Beaverlodge area in 1944. This led to the development of the first Saskatchewan uranium mine, the Nicholson mine, followed by the discovery of hundreds of other uranium ore deposits in the Beaverlodge area – many of which went on to be developed as uranium mines in their own right.

The Nicholson mine was the first uranium mine to be developed in Saskatchewan and, in 1949 was the only active uranium mine in Canada outside of the Northwest Territories. By 1959 the Nicholson ore body had been essentially depleted, however the Nicholson mine had played its role in helping Canada become one of the largest uranium producers in the world.

The Nicholson mine produced about 12,800 tonnes of uranium ore, yielding about 50 tonnes of uranium (as U_3O_8), and an estimated 60 to 90 thousand m^3 of waste rock. Following closure in 1960, the Nicholson site was abandoned with little remediation and no reclamation being done.

Forty-five years would pass before the governments of Saskatchewan and Canada reached an agreement to fund the remediation (clean-up) of the Nicholson site, and contracted the management of the project to the Saskatchewan Research Council (SRC). At the time of writing this book the front-end assessment and remediation were about to begin, with several years of clean-up activity anticipated, and then a subsequent monitoring period, before the site is expected to be released into a long-term management and monitoring program.

ACKNOWLEDGMENTS

Thanks to Ann Marie Schramm, Larry Evans, William Schramm, Ian Wilson, and Dr. Joe Muldoon for reading and commenting on drafts of this book.

Many thanks also to George Bihun, Saskatchewan Ministry of Environment, for his help finding historical documents and photographs, and to David Thomas and Cory Hughes (Saskatchewan Ministry of Energy and Resources), and Dwayne Pattison (SRC) for supplying additional information on Nicholson.

Even in the modern electronic and Internet age there remains a need for major research libraries with substantive collections of scientific, technical, and engineering books and periodicals. In the preparation of this book our work was greatly assisted by the collections of the libraries of the University of British Columbia, University of Saskatchewan, University of Regina, University of Calgary, and the University of Toronto.

1 THE URANIUM AGE

1.1 Uranium Exploration – When it All Began (1789-1937).

The mineral pitchblende was first identified and named in 1727 as the rock that glowed[1], at St. Joachimsthal in what is now Czechoslovakia [1]. This was before uranium itself was discovered by German scientist Martin (W.H.) Klaproth in 1789. At first, pitchblende was simply a curious by-product of silver mining, and it seems to have been known in various parts of Europe by the late 1700s. It was later found to be a uranium oxide mineral (also known as uraninite) whose average chemical formula is U_3O_8. Some limited mining of pitchblende for use in colouring glass and porcelain [2] took place in Europe in the early 1800s (and probably in earlier centuries as well), but there was relatively little interest in the mineral until the discovery of X-rays by German scientist Wilhelm Röntgen in 1895, radioactivity by French scientist Henri Becquerel in 1896, and radium in 1898 by French scientists Marie and Pierre Curie (see Reference [3]). Radiation therapy for various diseases, particularly cancer, emerged shortly after the discovery of X-rays, and radium therapy became an even more popular method for radiation treatments beginning in the early 1900s. At about the same time radium came into industrial use as a luminous coating for glow-in-the-dark products, particularly instrument panels, watches, and clocks.

In the United States (U.S.), uranium ore was discovered in 1871 in gold mines near Central City, Colorado and later at the Colorado Plateau of Utah and Colorado. These areas were actively mined for their vanadium and/or radium contents in the late 1800s and early 1900s [4,5].

---

[1] The glow was probably due to radiation from uranium and radium causing zinc sulfide, which is also found in this particular pitchblende mineral, to phosphoresce [1].

Industrial quantities of uranium, in the form of pitchblende, were first discovered at Shinkolobwe in the Belgian Congo[2] in 1915, as brightly coloured "queer stone[s]" [6]. There wasn't much industrial interest in uranium at that time but there was a market for radium[3], which is commonly associated with pitchblende. The Belgian company Union Minière du Haut Katanga mined the Shinkolobwe deposit for its radium, beginning in 1921 and continued for nearly two decades [1,6]. In the early years of radium therapy 100 mg of radium salt sold for about $12,000 (in 1918 U.S. dollars) [7]. When the market for radium severely weakened near the end of the 1930s, the Shinkolobwe mine was closed.

Canada's uranium story had begun somewhat earlier, when in 1900 two scientists from the Geological Survey of Canada, Mackintosh Bell and Charles Camsell, noticed brightly coloured *"lilac stain of cobalt"* on rocks along the shore of Great Bear Lake, Northwest Territories, [6,8]. Bell and Camsell didn't pursue their observations, but they did write a report on their findings. Nearly thirty years later prospector Gilbert LaBine studied their report as part of his research for upcoming prospecting expeditions. In the 1920's Gilbert and his older brother Charles LaBine had opened the Eldorado Mine in Manitoba, and they formed the Eldorado Gold Mines Ltd. in 1926 [9,10]. When the Eldorado Mine played-out, Gilbert LaBine returned to prospecting and began searching near Great Bear Lake [11]. In 1930, LaBine and his prospecting partner Charles St. Paul found Bell and Camsell's cobalt stains and discovered deposits of silver, cobalt, and pitchblende [6,8,9,12]. To develop this find they established a new Eldorado Mine near what would later become Port Radium[4] (see Figure 1.1). They began some mining in 1931/32 and brought the mine and a mill into continuous production in 1933 [8,10].

Uranium-bearing rock (pitchblende) was first discovered in Saskatchewan in 1935 at the Nicholson copper prospect on the north shore of the Athabasca [14] and also in several other locations near the north shore of the lake between 1935 and 1936 [8,15-17].

---

[2] The Belgian Congo achieved independence in 1960, and since then has been the Democratic Republic of the Congo.

[3] At this time radium was in demand for use in cancer treatments.

[4] There were several name changes, first it was called Great Bear, then Cameron's Point in 1932 [1]. By 1933, with prospecting and mining on the increase in this area, the Canadian Government established a town named Cameron Bay, about 11 km south-east of present-day Port Radium. The government facilities in Cameron Bay were renamed Port Radium in 1936 and eventually the whole community came to be referred-to as Port Radium.

Figure 1.1. Illustration of the location of Canada's first uranium mine - the new Eldorado mine and mill – near Port Radium, east of Great Bear Lake (1). Also shown is the location of Eldorado's uranium refinery at Port Hope on the north shore of Lake Ontario (2). The map itself approximates Canada as it was in the 1930s, and was drawn based on the *"Territorial Evolution, 1927"* map in the *Atlas of Canada*, 6th Ed. [13].

None of these were followed-up with significant additional exploration and discoveries until 1944 when Eldorado Mining and Refining staked several claims in the Beaverlodge area (see Figure 1.4). Others followed suit in subsequent years. In 1949 uranium exploration in the area had dramatically increased and the abandoned town of Goldfields was revived [15]. Although thousands of radioactive "surface showings" had been discovered by this time, none but Nicholson would be developed until the beginning of the cold-war in 1951. In addition to developing the new Eldorado Mine and the Port Hope refinery, LaBine continued to explore in central Manitoba where, in 1934, he formed Gunnar Gold Mines, which remained in production for several years [18]. As will be discussed below, a second Gunnar mining company was to come later – in the 1950s.

Demand for radium exceeded supply throughout the 1930s, keeping prices high and making radium mining and milling quite profitable. By 1940, however, the adverse effects of radium on human health had become well known, and its use in most medical treatments, consumer-health-products, and luminescent products had been discontinued. As a result, the radium market collapsed. This coupled with the existence of large inventories on hand caused the new Eldorado Mine to be closed in 1940 (as was the Shinkolobwe mine). The Port Radium mine was also virtually abandoned. However, both the Eldorado mine and refinery were destined to become revitalized only two years later, as a result of the international atomic energy race.

1.2 Nuclear Fission is Discovered – The Atomic Age Begins (1938-1941).

In 1938, two German chemists, Otto Hahn and Fritz Strassmann discovered that when they bombarded uranium nuclei the nuclei split apart, yielding two approximately equal fragments of a lighter element, barium, however the mass of the fragments totaled to less than that of the original uranium [3,19]. Seeking to explain this phenomenon, two Austrian physicists, Otto Frisch and Lise Meitner developed a very important theory. They theorized that the nucleus of a uranium atom, when struck by neutrons, could be split into pieces – the smaller atoms observed by Hahn and Strassmann – plus neutrons, and a huge amount of energy. This process became known as fission. They published their theories in 1939 [20].

Subsequently Hahn, Meitner, and other physicists realized that such fission actually occurred for a specific isotope of uranium (U-235) that was normally present only in very low concentrations, typically less than one percent. However, if enough U-235 atoms were packed closely enough together, then the neutrons released by one atom could cause the breaking (or fission) of several other uranium atoms, which in turn would split apart releasing more neutrons, and so on, creating a very fast chain reaction and releasing an extraordinary amount of energy [3,21]. The potential to create such huge quantities of energy from very small amounts of material was both very exciting and very frightening.

While excited about the potential for a new source of almost unlimited energy, Frisch and Meitner immediately realized that such power could cause harm as well. Frisch wrote to the British government warning that a small piece of uranium could *"produce a temperature comparable to that of the interior of the sun. The blast from such an explosion would destroy life in a wide area … probably cover[ing] the center of a big city"* [22]. Scientists in other countries, particularly in Germany, Russia, Canada [3,23][5], and the U.S. came to similar conclusions, and thus began a race to find and obtain uranium and to try to develop atomic weapons.

It had been estimated that the critical mass needed to enable the chain reaction in an atomic bomb would be about 10 kg (22 lb) of U-235, although the first-ever atomic bomb actually contained 64 kg (141 lb) [22]. Countries involved in the atomic weapons race naturally desired to stockpile as much uranium as possible as a strategic resource for their own use. For a while it was thought that uranium was somewhat rare, with only three substantial deposits being known: Shinkolobwe in the Belgian Congo, Port Radium in Canada (see Figure 1.2), and St. Joachimsthal in a part of Czechoslovakia that had recently been annexed by Germany. The onset of the Second World War in 1939 heightened both strategic interests and fears, as it was also thought that the number of nuclear weapons could be limited by acquiring as much of the known uranium reserves as possible.

Figure 1.2. Photograph of Port Radium in the 1930s. (Courtesy: Public Archives of Canada # C-23966).

---

[5] In Canada, for example, Ernest Rutherford noted in 1904 that "the total energy emitted from 1 gram of radium during its changes is about a million times greater than that involved in any known molecular change … There is thus reason to believe that an enormous store of energy could be obtained from a small quantity of matter" [23].

These factors set the stage for renewed uranium exploration and atomic power developments in Canada and elsewhere. The National Research Council built Canada's first laboratory-scale fission reactor in Ottawa in 1940 using uranium from Port Hope, via the Eldorado mine that had just closed [24].

1.3 Substantial Uranium Resources Needed – The Atomic Age Reaches Canada (1942-1950).

Canadian exploration for uranium surged again beginning in 1942, due to military interest in building an atomic weapon. Eldorado had been selling their stockpiled, by-product uranium to the U.S. government in 1941 and 1942, but demand kept increasing [10]. In hopes of meeting the increased demand the Eldorado Mine was reopened in 1942 [24,25,26] and contracted to supply uranium to the U.S. Army, but it was clear that additional uranium reserves would also need to be found [18]. The Eldorado refinery at Port Hope had remained in continuous production, mostly producing radium, but by 1942 its focus had shifted to producing mostly uranium [1].

The Canadian government had imposed a ban on public prospecting for, and mining of, any kind of radioactive materials across Canada, but the government itself vigorously pursued these activities in secret [1,27][6].

In 1942 the United Kingdom (U.K.) and Canadian governments launched a joint atomic energy research program based in Canada [1]. By August 1943 the U.S., U.K., and Canada had merged their atomic weapons development programs under a cooperation agreement called *"The Articles of Agreement on Tube Alloys"* ("Tube Alloys" having been the code name for this project) [18,24]. This agreement was reached and signed in Canada at "The Quebec Conference," held at the Citadel in Quebec City. The "Tube Alloy" Project later became part of the Manhattan Project, led by the U.S. but still in cooperation with the U.K. and Canada. The world's first nuclear fission reactors were built in the U.S. and Canada. Part of Canada's role was to develop and build a heavy water reactor for the production of plutonium from uranium [24].

Included in the Manhattan Project was a component aimed at finding and acquiring as much uranium as possible, beginning in about 1942. At first, the uranium for the Manhattan Project came from the Shinkolobwe Mine in the Belgian Congo and from the Eldorado Mine (refined at Port Hope) in Canada. To this was later added uranium produced in the U.S. itself. Several American companies had been mining for vanadium in Colorado and Utah, but their ore actually contained both vanadium and

---

[6] Attempts to keep Canada's uranium mining secret were not entirely successful. In July of 1943 Eldorado received an order for uranium from the USSR [24].

uranium. The uranium in their ore had now become valuable, but to maintain secrecy the U.S. Army publicly maintained that they were only buying vanadium. At this point in time the largest known deposit of uranium was still at Shinkolobwe, but there was now a strategic reason to search worldwide for additional reserves: a desire on the part of the U.S., U.K., and Canada to control as much as possible of the world's uranium reserves. This was later enhanced by a growing interest in plutonium (made from uranium).

In 1942, the Canadian Government passed legislation reserving to the Crown ownership of all radioactive substances found in the Northwest Territories and Yukon. In order to maintain security the Canadian government also started purchasing shares in the Eldorado company [10,24]. In 1943 Eldorado Gold Mines Ltd. was renamed Eldorado Mining and Refining Ltd.[7] [1], and in 1944 the Canadian government expropriated all outstanding shares in the company and turned it into a Crown Corporation. Gilbert LaBine remained President of Eldorado through all of this, and had shifted the Eldorado Mine and Port Hope refinery focus to the production of uranium rather than radium. This involved converting uranium ore concentrate (yellowcake) into uranium black oxide (an orange-coloured solid comprising about 96% U_3O_8) [28]. For the next several years, this mine and refinery were the only significant new source of uranium in the western world. Also during this period, the Port Hope refinery received and processed ore from the Shinkolobwe mine [10]. Most of the uranium was sent to the U.S. and probably used to produce the first atomic bombs[8], including the first plutonium bomb ("Trinity") detonated at Jornada de Muerto, New Mexico in July, 1945 [9].

In 1946 the Canadian government established the Atomic Energy Control Board (AECB) to regulate the uranium industry and essentially all atomic energy activities [24]. In 1947 Canada began using Eldorado to stockpile uranium, in addition to supplying the U.S. [10]. Meanwhile, the Canadian government pursued additional secret uranium exploration and development activities across Canada using the resources of both Eldorado and the Geological Survey of Canada [12,27]. This wave of uranium exploration was aided by the availability of hand-held portable radiation detectors[9] [29,30]. The most significant finding of this relatively new wave

---

[7] Eldorado Mining and Refining Ltd. later became Eldorado Nuclear Ltd.

[8] It has been estimated that about one-sixth of the uranium delivered to the US Manhattan Project came from Canada [5].

[9] Some of the first commercial hand-held radiation detectors include ionization-chamber detectors, such as the 1930s-era "Curtiss Radium Detector" and the 1940s-era "Victoreen Model 247/247A," and Geiger-Müller counters, such as the 1930s-era "Radium Hound" [29,30].

was that the Beaverlodge region north of Lake Athabasca in northern Saskatchewan not only had pitchblende (as had been previously discovered in 1934) but had at least a thousand pitchblende occurrences [1]! Of these, the first staking took place in 1944 [31] and the first large ore body was discovered in 1946 [1].

When the Second World War-era security restrictions were reduced in 1948 [6,12] private enterprises were again allowed to get involved in exploration, mining, and milling, although all mined ores and concentrates were still required to be sold to Eldorado or other government-designated agency (and at a government-guaranteed price) [1,8,10,12,32]. The renewed exploration activities of 1948 and 1949 catalyzed the finding and development of new uranium deposits and about 45 small- to medium-size mines [33,34] (see Table 1.1). An example is the Madawaska/Faraday Mine near Bancroft, Ontario, which was discovered in 1949 but did not begin operating until 1957 [35]. The first Saskatchewan uranium mine to be developed was the Nicholson Mine, which commenced production in 1949 [16].

In 1949 the only Western suppliers of uranium were just Eldorado, Nicholson, and Shinkolobwe but by 1950 both the U.S. and South Africa had also become significant producers of uranium. The price for uranium concentrate guaranteed by the Canadian government was increased in 1950 to encourage further exploration and the development of additional new uranium mines and mills [32].

The end of this era was also the beginning of the Cold War era. The pursuit of military and peaceful applications of nuclear energy, driven by both hope and fear, rekindled demand for uranium. By the end of the 1940s Russia was receiving all of the uranium ore produced by the St. Joachimsthal mine in Czechoslovakia, and by the Schlema mine in East Germany, and it had become known that Russia had successfully test-exploded an underground atomic bomb.

Table 1.1. Examples of Canadian Cold War-Era Uranium Mines.
(Sources: [16,17,33-38]).

Mine	Discovery	Mean Grade (% U)	Producing Years	Yield (tonnes U_3O_8)
Beaverlodge, SK (Eldorado Ace-Fay-Verna Mines and Mill)	1946	0.24	1953 – 1982	~20,400 [35,36]
Cayzor Athabasca Mine, SK	~1953	0.33	1954-1960	221 [16]
Cinch Lake Mine, SK (later Lake Cinch)	1948	0.20	1955-1960	336 [16]
Eldorado – Dubyna Mine, SK	1947	0.22	1978-1982	192 [16]
Eldorado – Eagle Mine, SK	~1946	Erratic	1950-1951	100 [16]
Eldorado – Fish Hook Mine, SK	1945	0.22	1957-1960	18 [16]
Eldorado – Hab Mine, SK	1958	0.43	1972-1976	900 [16]
Eldorado – Martin Lake, SK	1946	Erratic	1948-1954	13 [16]
Eldorado Port Radium Mine and Mill, NWT	1930		1930–1940; 1942-1960	
Gunnar Mine and Mill, SK	1952	0.18	1955 – 1963	8,133 [16]
Lacnor Mine, ON	1953		1957-1960	2.7 million
Lorado Uranium Mine and Mill, SK	~1953		1956-1960	105 [16]
Madawaska/Faraday Mine, ON	1949		1957-1964; 1975-1982	4,305
National Explorations - Keiller , Pat Mines, SK	1951	~0.5-0.8	1954-1958	35 [16]
Nesbitt-Labine Uranium – Eagle, ABC Mines, SK	1950, 1952	0.15-0.24	1952-1956	27 [16]
Nicholson Mine, SK	1935	0.3 – 0.5	1949-1959	48 [16]
Pronto Mine and Mill, ON	1953		1955-1960	2.1 million
Rayrock Mine, NWT	1948		1957-1959	207 [36]
Rix Athabasca – Leonard Mine, SK	1951	~0.2	1955-1960	91 [16]
Rix Athabasca – Smitty Mine, SK	~1949		1952-1960	514 [16]
Uranium Ridges Mine, SK	1950	0.53-0.75	1958-1959	12 [16]

Canada had developed a series of research reactors during this era. The Zero-Energy Experimental Pile (ZEEP) Reactor was Canada's first nuclear reactor and the world's first non-U.S. reactor. It operated from 1945 to 1970 and was used to produce plutonium and uranium-233. Canada's second nuclear reactor, the National Research Experimental (NRX) Reactor, commenced operation in 1947 and remained in service until 1993. Meanwhile Eldorado had begun to sell cobalt-60, and by 1951 Eldorado, two Canadian university groups (in Saskatchewan and Ontario), and groups in the U.S. had developed cobalt-60 medical devices (called "cobalt bombs") to provide focused gamma rays[10] for radiation treatment of cancer [10,39]. By the early 1950s Canada had become the world's largest supplier of medical isotopes [40].

1.4 The Cold War-Era Uranium Mines (1951 – 1967).

With the beginning of the cold war the U.S. decided to continue its nuclear program, including expanding their nuclear arsenal and conducting research and development (R&D) aimed at developing a hydrogen bomb. These activities increased the demand for uranium from Canadian and U.S. mines. Russia had also decided to continue with its nuclear program, drawing uranium from Schlema in Germany and St. Joachimsthal in Czechoslovakia. The early 1950s also saw Britain independently continue its nuclear program, drawing uranium from Portugal, the Belgian Congo, South Africa, and France launched a nuclear weapons development program as well. In Canada at this time, the Port Hope refinery became exclusively focused on producing uranium [1].

Beyond uranium-security and weapons programs, another factor contributing to demand for uranium was the emergence of nuclear power programs[11]. The United States had started a nuclear power program in the 1940s, and the first electric-power generating nuclear reactor, EBR-I[12], was built in Idaho and started-up in December, 1951 [22,41]. For its part, Canada had launched Atomic Energy of Canada Ltd. (AECL, a Crown Corporation) in 1952 and a nuclear power program in 1955, producing the nuclear power demonstration (NPD[13]) reactor, which was built in Ontario

---

[10] Gamma rays are electromagnetic waves of very high energy (and very short wavelength). Gamma rays are one of the kinds of radiation that can be produced by radioactive atoms as they decay.

[11] Industrial applications were also being developed, such as in nuclear density gauges and nuclear thickness gauges, but these did not significantly affect overall uranium demand.

[12] The EBR-I (Experimental Breeder Reactor I) produced 200kW of electricity and was operated from 1951 until decommissioning in 1964.

[13] NPD was the fore-runner of the Canada deuterium uranium (CANDU) power reactors.

and started-up in June, 1962 [10,42]. The emergence of these nuclear power programs contributed to governments' desire to build uranium reserves, while regulatory and incentive changes by the Canadian and U.S. governments triggered uranium exploration rushes in both countries [1,4,10].

The regulated (Canadian) price for uranium was increased in 1950 and again in 1951 to stimulate exploration and to ensure that the Beaverlodge and other mines could proceed [32]. The Beaverlodge mine did proceed (beginning operations in 1953), as did a number of other, smaller mines [1,10]. In response to the "tent cities' that had begun to spring up around the individual mine sites, the Saskatchewan government established the community of Uranium City in 1951 with the aim of serving the entire region (see references [37,43-45]). Then, in 1952, the Saskatchewan government changed regulations making it more attractive for prospectors to explore and stake claims in the Lake Athabasca region (for which the Canadian Government would have exclusive purchase rights for all mined uranium) and to support them [44]. These steps made it attractive for more companies, and even amateurs to prospect for uranium, triggering a massive uranium exploration and claim-staking rush [4,44,46-50] that helped Canada maintain its international position as a uranium producer[14]. A *Northern Miner* headline proclaimed "Uranium - Canada Maintains Place in Frantic World Production Race" [51] and a *Precambrian* article noted that *"Uranium deposits, it seemed, began to appear everywhere"* [52]. The Saskatchewan uranium rush even caught the attention of broadly circulated magazines like *Maclean's* and *Life* [53-56], and made headline news as far away as Australia [57-60] (see Figure 1.3). A television documentary film, *"The Birth of a Great Uranium Area,"* was made in 1953, illustrating the processes of uranium prospecting, drilling, and mining in the area [61]. By the fall of 1954 the government announced that 50 to 60 companies were actively engaged in uranium exploration, development, mining and/or processing in Northern Saskatchewan [62]. Uranium City itself grew to nearly 5,000 people (the size for which it had originally been designed [45]), see Table 2.1 below. Another television documentary film, *"The Road to Uranium,"* was made in 1957, illustrating life in Uranium City at its peak of 5,000 people, and of operations at the Eldorado mines and mill [63].

---

[14] At about the same time, related developments in the United States triggered a uranium rush there as well [4].

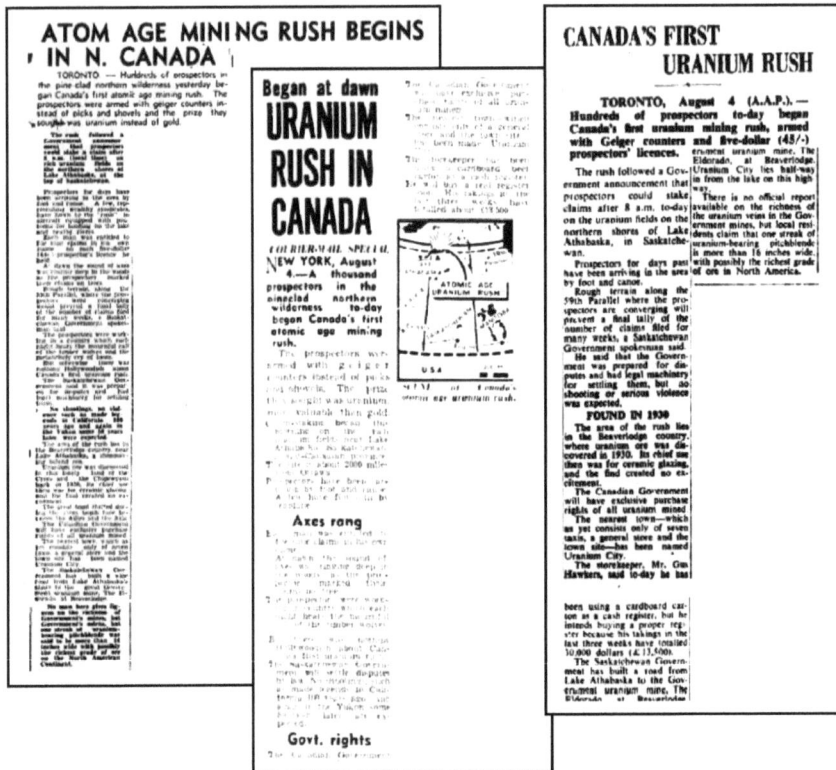

Figure 1.3. Illustration of Australian newspaper headlines reporting the 1952 Saskatchewan uranium exploration boom [57-59].

The increased level of exploration activity driven by the Cold War led to several significant Canadian uranium discoveries[15] beyond those in Saskatchewan, including deposits around the Bancroft, Ontario, area in the early 1950s (including the Madawaska/Faraday Mine mentioned earlier), and the first discovery in the Elliot Lake / Blind River, Ontario, region in 1953. This led to 12 mines including the Lacnor, Milliken, Nordic (Algom Nordic), Panel, Pronto, Quirke (Algom Quirke), Stanleigh, Stanrock, Can-Met, and Denison Mines [6,27,38] - see Table 1.1 and Figure 1.4.

---

[15] The search for additional uranium resources was not restricted to Canada, of course, and significant new deposits were found elsewhere during this period, such as at Mi Vida in Colorado in 1953.

Before World War II the cutoff grade for commercial production of uranium was about 0.5% (as U_3O_8), but the ability to apply more sophisticated processing technologies, and to work at larger scales of operation, brought the cutoff grade down to as low as 0.1 % by the mid-1950s [6]. This change enabled many new mines to be developed in Canada and, again by the mid-1950s, several practical deposits had been identified [27] although some either did not get developed or did not last very long.

For example:

- In 1949 uranium mineralization was discovered to be associated with silver and lead minerals in the Rexspar deposit in southern British Columbia. The deposit was further evaluated throughout the 1950s (and again in the 1960s and '70s) but was never brought into production [64].
- The Rayrock Mine near Great Slave Lake, Northwest Territories (see Figure 1.4) was discovered in 1948. Production at Rayrock began in 1957, but the mine only lasted for two years [1,35]. The mine and its small town-site were both abandoned in 1959 when its reserves were depleted [1,35].

There had been pitchblende prospecting and staking in Northern Saskatchewan since the 1930s and 1940s, but by 1952 there was enough uranium prospecting activity that the Saskatchewan government established the town of Uranium City, as noted above, to provide services to the mines in the Beaverlodge area north of Lake Athabasca (see Figures 1.4 and 1.7). Several Saskatchewan discoveries followed that of Beaverlodge, including Radiore nearby, Rix Athabasca on Black Bay, and Nesbitt-Labine near Uranium City [10]. Gilbert LaBine and his son Joseph were part of the renewed prospecting in this region, and they found several deposits in the same general area in 1952 [6,9].

In July 1952 LaBine, Albert Zemel, and Walter Blair found and staked a uranium deposit at the southern tip of the Crackingstone Peninsula on Lake Athabasca [1,65] (see Figure 1.5). This became the site of the Gunnar Mine, which was the richest uranium strike in Canada at the time and would become the first large private uranium mine of the era [10,56]. When the Gunnar Mine and mill site opened in the fall of 1955, it doubled Canada's uranium production capacity [66-68] (Figure 1.5).

Figure 1.4. Illustration of the locations of some of Canada's cold war era uranium mines: Eldorado Mine (1), Bancroft-area mines (2), and Elliot Lake - Blind River area mines (3), Rayrock Mine (4), Beaverlodge-area mines (5). The map itself approximates Canada as it was in the 1950s, and was drawn based on the "*Territorial Evolution, 1949*" map in the *Atlas of Canada*, 6th Ed. [13].

Through Eldorado Nuclear the LaBines began mining at their Beaverlodge site[16] in 1953, where they established a dedicated mill and also a small community named Eldorado, all located about 7 km east of Uranium City [27,36]. This site, which included the Ace, Fay, and Verna mines[17], operated until 1982 and was the highest producer of the 16 Beaverlodge area mines of the Atomic Age and the Cold War Eras (Table 1.1; [67]).

---

[16] The Beaverlodge Mine has sometimes been referred-to as the Eldorado Mine, although as noted above there were two previous Eldorado Mines, one in Ontario and an earlier one in Manitoba.

[17] An illustration of the early days of establishing Eldorado's Ace and Fay mines is given in a TMC documentary [61].

Figure 1.5. The Gunnar open pit mine in 1959 [69].

The Lorado Mine opened in 1957 and operated until 1960 [35] (see Figures 1.6, 1.7). Lorado was the only other mine to have its own dedicated mill (along with Beaverlodge and Gunnar), and the Lorado mill operated from 1957 to 1961. In addition to processing the mined ore from their own mines, the Lorado and Beaverlodge mills also processed ore from smaller mines in the region, including the Cayzor, Rix Leonard, and Cinch Lake Mines, and also ore from "surface miners" who picked over surface showings and waste rock piles that were uneconomic for the mining companies to handle [35,65].

In total, more than 1,000 pitchblende occurrences were discovered in Saskatchewan's Beaverlodge district, however only 16 mines of significant size were actually brought into production between 1953 and 1982 (see Table 1.1) [16]. By 1960 LaBine estimated that the Beaverlodge district mines had produced about $300 million worth of uranium (in 1956 dollars) [31].

Figure 1.6. Lorado uranium mill site in 1958 [70].

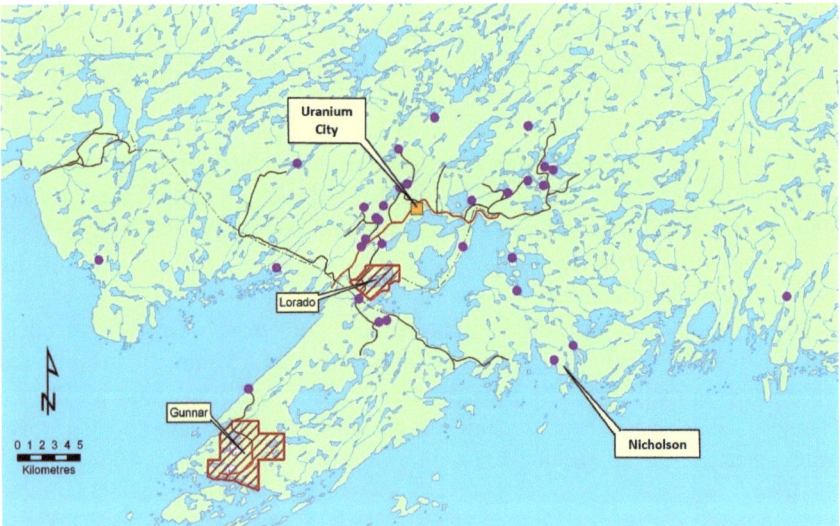

Figure 1.7. Illustration of the locations of three of Saskatchewan's cold war era uranium mines: the Gunnar, Lorado, and Nicholson Mines.

By 1957 there were 18 operating uranium mines in Canada (see Table 1.1), all of which sold their uranium ore (raw or milled) to Eldorado, which in turn sold the uranium to the U.S. [10]. This number peaked at 21 Canadian producing mines in 1958, by which time 11 of the mines plus three mills were operating in the Uranium City area [1,12,35].

Canadian uranium production levels grew throughout the 1950s. In 1956 the "free-world" production of uranium was close to 13,000 tonnes [6]. Most of the production came from the U.S., Belgian Congo, and Canada (see Table 1.2). By 1958 half of the world's production of about 27,200 tonnes was estimated to have come from "Arctic Canada" meaning the Port Radium Eldorado Mine [6,27]. By 1959 uranium was Canada's number one mineral export (ahead of aluminum, iron, and nickel) with 23 mines and 19 mills in operation [27]. Of the 19 mills, 11 were in the Elliot Lake, Ontario area, three near Bancroft, Ontario, three in northern Saskatchewan, and two in the Northwest Territories [27].

Table 1.2. World uranium reserves in 1957.
(Conversions to S.I. units are approximate. Based on data in Reference [6].)

Country	Production in 1956 (tonnes U_3O_8)	Uranium Reserves (millions of tonnes U_3O_8)	Average Ore Grade (mass%)
South Africa	3,992	998	0.03
Canada	2,994	204	0.10
United States	5,443	54	0.24
Total	12,429	1,256	

At about this time radiation safety was becoming better understood. Although by 1940 the adverse effects of radium on human health had become well known and its use in most medical treatments and consumer products had been discontinued, the health effects of low doses of radiation were not yet well known. The concepts of safe working levels and safe cumulative (annual) exposure levels emerged in the 1950s, although uranium mining in the 1950s was still considered "safe" (as far as radiation hazards were concerned) [24,71]. In 1959 Gordon Churchill, the federal Minister responsible, confidently stated in the House of Commons that "... *there are no special hazards attached to the mining of uranium that differ from other mining activities"* and *"... there is no radiation hazard in the processing operations"* [24]. Nevertheless, in 1960 the AECB created regulations dealing with radiation safety [24]. The 1960 AECB regulations defined for the first time the concept of an *"atomic energy worker"* and the maximum amounts of ionizing radiation to which such a worker could be allowed to become exposed[18].

Most (about 90%) of the Canadian uranium produced in the 1950s and early 1960s was sold to the U.S. Atomic Energy Commission, with most of the remainder being sold to the U.K. Atomic Energy Authority [1,8,12]. The Rayrock mine closed in 1959. Then the Eldorado Mine closed in 1960, when the uranium ore ran out, although the Eldorado mill was able to continue operating until 1967. Despite depleting ore bodies, by this time uranium production was outstripping demand. By 1962-63 the U.S. had more than enough uranium for its needs and the U.S. Atomic Energy Commission began reducing its purchases [29]. As a result, the levels of both exploration and mining and milling decreased, and the number of active mines shrunk to only five, the Madawaska/Faraday, Milliken, and Nordic mines in Ontario, and the Beaverlodge and Gunnar mines in Saskatchewan [12,27]. A 1966 Saskatchewan Department of Mineral Resources report concluded that *"uranium was a glut on the market, and without markets incentive is lacking to search for and develop new mines"* [12]. For a while the Canadian government supported the uranium industry with a stockpiling program, but this only lasted until 1974 [27].

Although there was an oversupply of uranium in the markets, the late 1960s were also characterized by much eager anticipation of nuclear power and clean energy. As already noted, first power-generating nuclear reactor, EBR-I, had begun operating in Idaho, but not many other power reactors had yet been built. Nevertheless, with forecasts that the world's known reserves of uranium[19] could be quickly used-up if many nuclear power plants were constructed [12], prospecting for new deposits increased again, and once again Saskatchewan was *"now one of Canada's busiest [uranium] prospecting areas"* [12].

Uranium production in the Bancroft and Beaverlodge areas ended in 1982, and in the Elliot Lake area in 1996 [27]. Thus ended the second era of Canadian uranium production. The first two eras of Canadian uranium production helped put Canada on the world stage from an industrial point of view, and these eras are rich in stories. One of these is the story of the

---

[18] These regulations also defined the maximum amounts of ionizing radiation to which a member of the general public could be allowed to become exposed, at 1/10th of the amount for an atomic energy worker.

[19] In 1964 the known world reserves of uranium were 430,000 tonnes, 40% of which was located in Canada (Saskatchewan Department of Mineral Resources, 1966).

Nicholson Mine, which is described beginning in Chapter 2[20].

The Beaverlodge area mines and smaller developments of the Atomic Age and the Cold War Eras left behind an unfortunate legacy in that they were simply abandoned without much or any cleanup, and frequently without significantly closing-off the various mine shafts, adits, and raises. A 2006 Saskatchewan Environment and Resource Management report identified 45 such abandoned mines in the immediate Uranium City area alone [39]. These aspects are described further, in the context of the Nicholson operations, in Chapters 6 and 7.

There was a significant lull in Canada's uranium production beginning with the end of the cold-war era in about 1967. This was partly due to market conditions, as described above, but also partly due to changes in the political and regulatory environment. Although the Atomic Age and the Cold War Eras (1938 through 1967) exhibited uranium mine developments across Ontario, the Northwest Territories, and Saskatchewan (see Table 1.1), the modern-era uranium developments in Canada would all take place in Saskatchewan. (Saskatchewan also has one nuclear reactor in operation. The Saskatchewan Research Council has operated a SLOWPOKE 2 research reactor in Saskatoon since 1981 [72].) Between 1982 and about 1992, the Saskatchewan government phased-out uranium mines and mills, however from about 1992 onwards the province reversed course and once again became an active supporter and co-developer of uranium mining in Saskatchewan [73]. Whereas the early Saskatchewan developments were in the Beaverlodge area immediately north of Lake Athabasca, the next boom in uranium exploration (in the 1970s) resulted in huge uranium discoveries in the Athabasca Basin area immediately south of Lake Athabasca. These deposits were not only large, but were of extremely high grade, enabling a huge increase in Canadian uranium production. From about 1992 through 2008 Saskatchewan became the uranium capital of the world with the highest productions levels and the largest deposits of the highest grade of uranium on the planet. By 2008 more than ten times as much uranium had been produced from the Athabasca region as was produced during the entire operating history of the Beaverlodge region (322 thousand tonnes vs. 30 thousand tonnes as U_3O_8) [64]. All of Canada's modern-day uranium production comes from mines in northern Saskatchewan (see Figure 1.8) [74] and has been about 12,900 tonnes of uranium per year from the late 1990s to the present [75].

---

[20] Another is the story of the Gunnar Mine, which is described in reference [68].

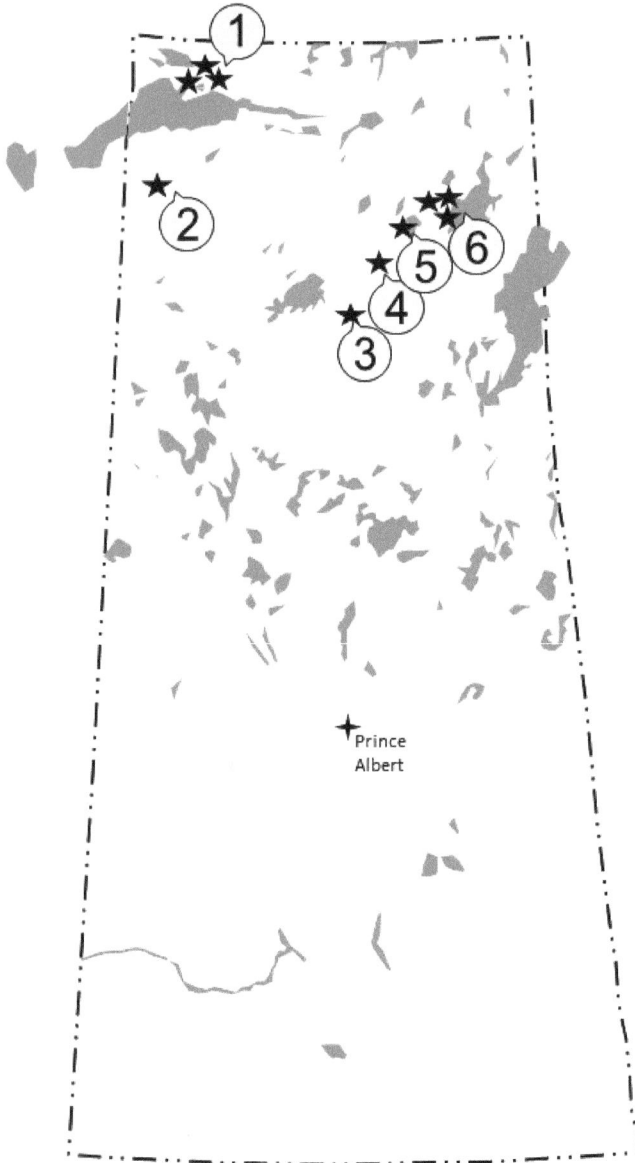

Figure 1.8. Illustration of the locations of Saskatchewan's cold war- and modern-era uranium Mines: Nicholson, Gunnar, Lorado, and Beaverlodge mines (1); Cluff Lake mine (2); Key Lake mine (3); McArthur River mine (4); Cigar Lake mine (5); and Midwest (proposed), McClean Lake, and Rabbit Lake mines (6). The city of Prince Albert is shown for reference. Drawn based on the "*Saskatchewan*" map in the *Atlas of Canada*, 6th Ed. [13].

2 THE BEGINNINGS AT NICHOLSON

As noted in Chapter 1, there had been pitchblende prospecting and staking in Northern Saskatchewan in the 1930s, 1940s, and between 1950 and 1952 a "uranium rush" had developed in Saskatchewan. The Nicholson mine history, however, began much earlier.

The general location of the Nicholson site is the northern shore of Lake Athabasca (Figures 1.4, 1.7, and 2.1). While this site is "near" Uranium City, it is actually approximately 16 km to the southeast and is only accessible by boat or float plane (from early summer through fall), and by vehicle or snowmobile over the ice (from winter through early spring).

The specific location is on the north side of a small bay that lies between Cornwall Bay and Fish Hook Bay (Figures 2.2 and 2.3). The 1936 and 1940 Geological Survey of Canada maps of the area do not show it as being named, and descriptions of it at the time instead described it in general terms such as being east of Goldfields, or "the area between Cornwall bay and Fish Hook bay" [14]. The site is not named on Robinson's Geological Survey map of 1950 [76], however, it is clearly identified as Nicholson Bay in the Geological Survey of Canada maps of 1949 [77], 1952 [78,79], and 1956 [80] therefore it seems likely that it was named "Nicholson Bay" in the late 1940s.

The Nicholson Mine property is usually described in terms of four main exploration zones, as shown in Table 2.1 and Figure 2.4. Some literature refers to a zone No. 6 just to the northeast of zone No. 1, where a small trench had been dug and small amounts of uranium and silver found. The various exploration sequences eventually revealed a high-grade uranium deposit in the No. 4 zone.

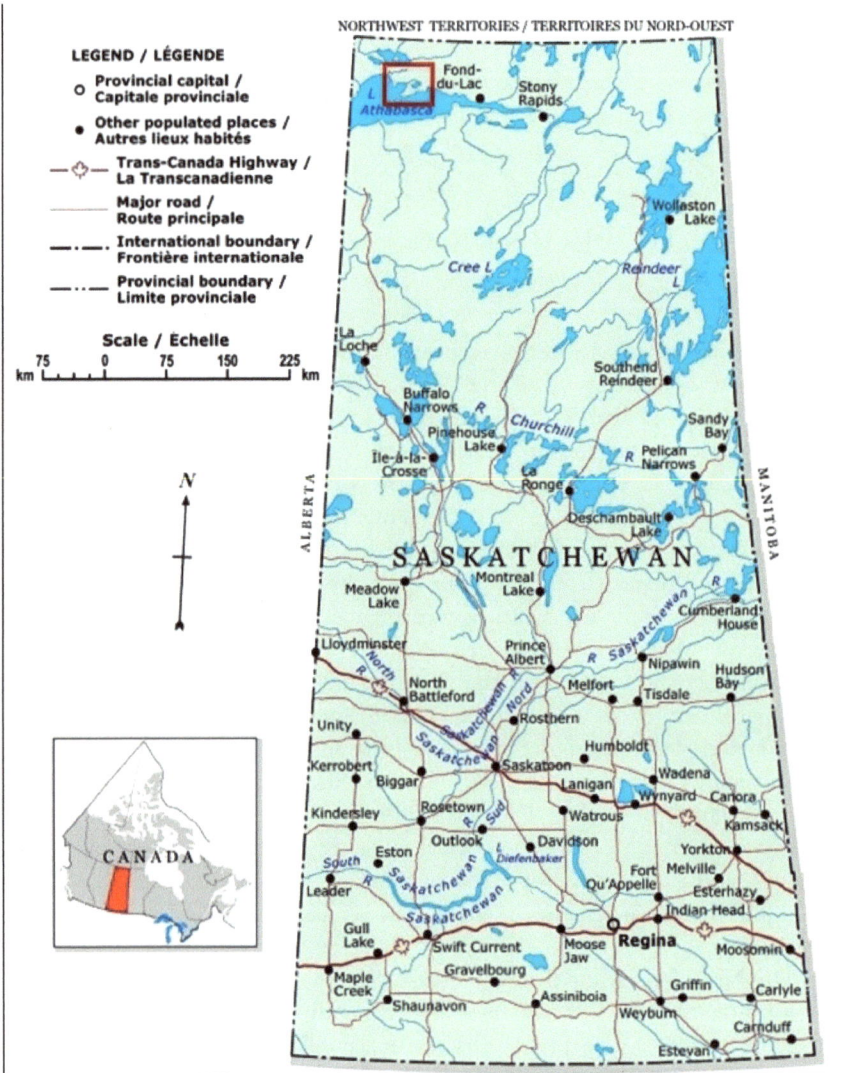

Figure 2.1. Nicholson site location (red square in extreme upper left corner of the map. (Map from Natural Resources Canada, 2001.)

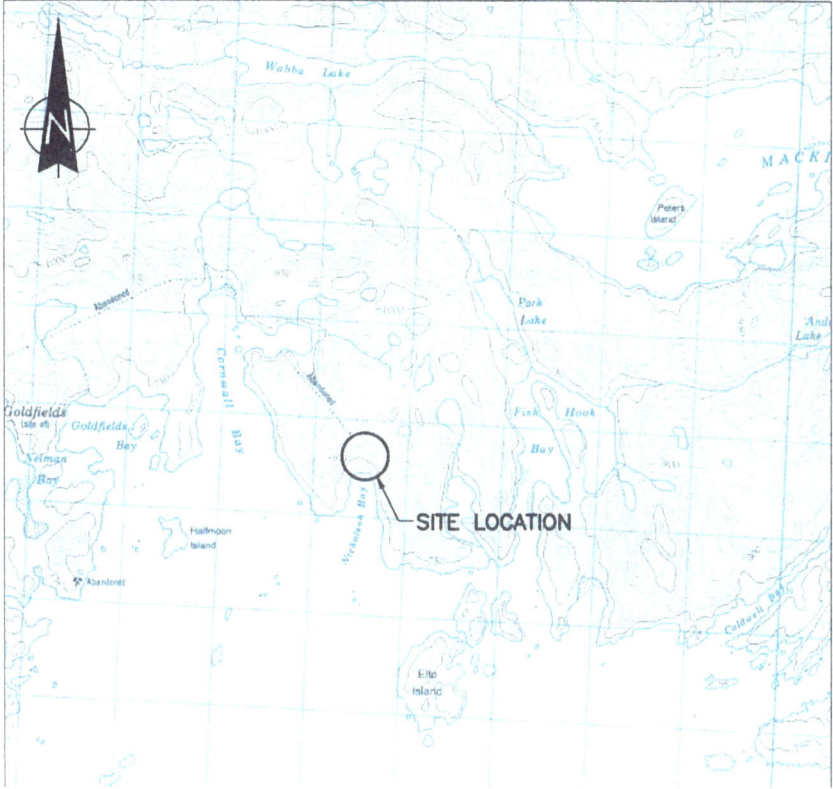

Figure 2.2. Nicholson mine area on Nicholson Bay. Note the abandoned road denoted to the northwest, and the Goldfields Bay area to the west. Reference [81].

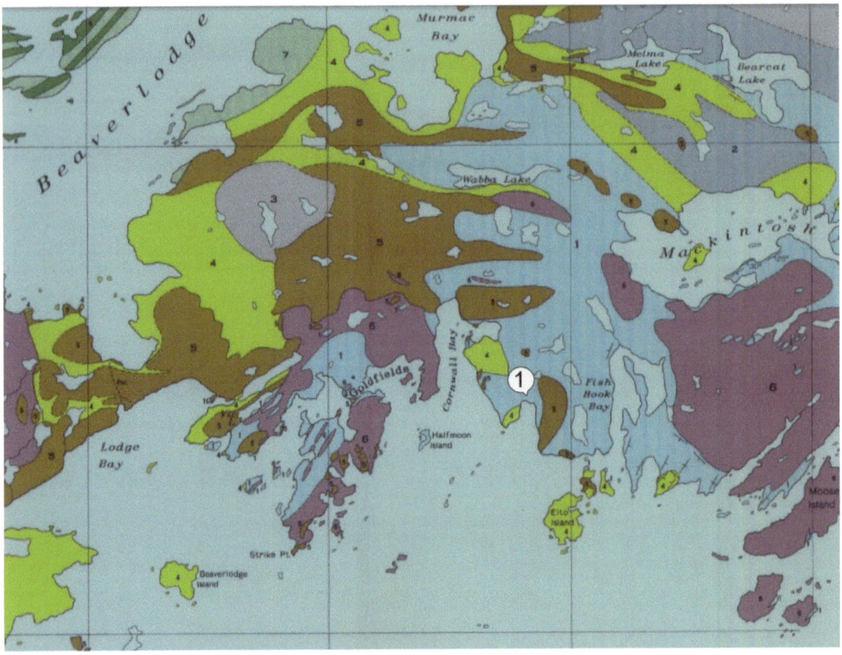

Figure 2.3. 1936 Geological Survey of Canada map showing the Nicholson site (1) near what would later become known as Nicholson Bay (between between Cornwall Bay and Fish Hook Bay). From Geological Survey Map 339A, "Goldfields Area," 1936, reference [14].

Table 2.1. Exploration Zones on the Nicholson Site. References [81,82,83].

Zone	Development and Location	Description
1	Exploration shaft. Southeast corner of the site, on the JIM claim.	Narrow zone about 60 m in length, consisting of carbonate, pitchblende, chalcopyrite, niccolite, cobalt-nickel, arsenides, and gold. A nearby extension was found, consisting of quartizite, containing iron, uranium, gold, and silver.
2	Exploration adit, shaft, and raise. Southern tip of the site near the Nicholson Bay shoreline.	Zone of about 55 m in length, consisting of quartzite, pitchblende, and other minerals containing some silver, gold, and platinum. A nearby extension was found, containing some strongly radioactive regions.
3	Undeveloped. Near the centre of the site.	A minor zone with a dolomite lens with mineralization that includes pitchblende. This was the site of early copper exploration, and silver has also been reported.
4	Main mine shaft, 3 adits (upper, lower and "New"), 1 raise, and stope breakthroughs. Northwest corner of the site.	The main zone having a length of about 490 m, and several offshoots.

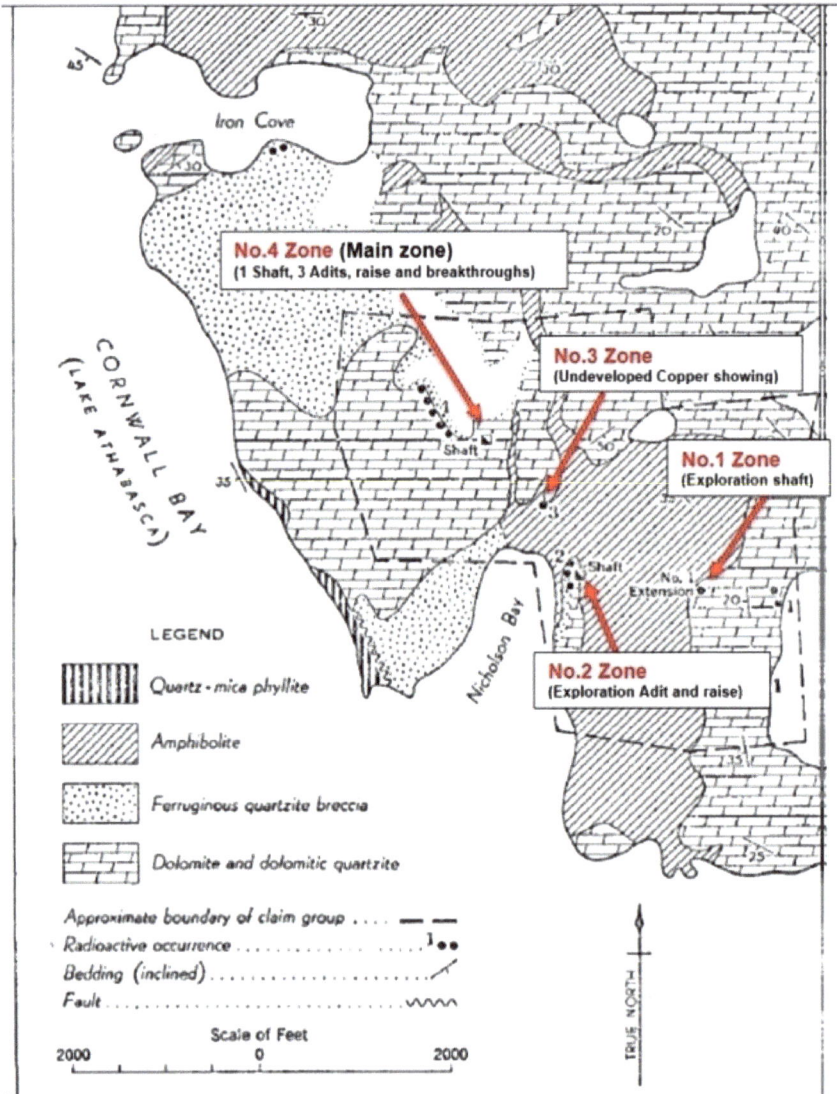

Figure 2.4. Annotated geological map showing the four principal exploration zones on the Nicholson site. Adapted from a 1962 Geological Survey of Canada map in Reference [83].

26

2.1 Prospecting for Iron.

As part of his exploratory survey of the Lake Athabasca area in 1893, Joseph B. Tyrrell noted the occurrence of iron-rich hills and sediments at Fish Hook Bay [15,84]. The nearby "Nicholson" area on the shore of the small bay between Cornwall Bay and Fish Hook Bay was first prospected and staked for iron in 1920 [81,85]. The original claims were named FLO, JIM, IVEY, DOROTHY, PETER, MARGARET, APEX, and STORM [81,85], see Figure 2.5. Other prospectors (E.A. and N.C Butterfield) discovered and staked five claims for iron in the immediate vicinity of Fish Hook Bay itself in 1921 [86,87].

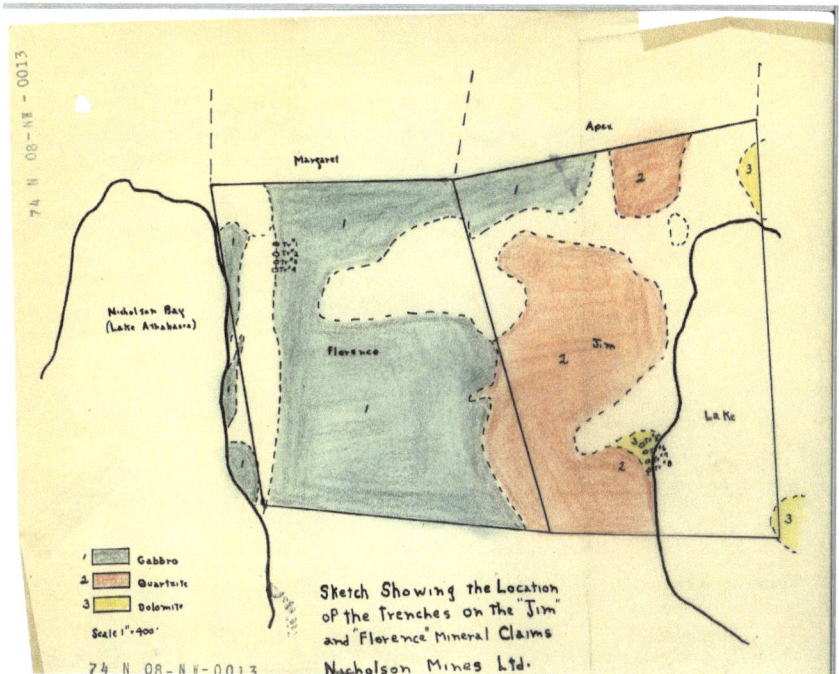

Figure 2.5. Surface plan showing some of the original claims on the site (Nicholson Mines Ltd., circa 1950).

Unfortunately, work by the Geological Survey of Canada in 1922 and 1946 showed this, and other iron deposits in the area of Fish Hook Bay, were below commercial grade and not worth developing [88,89]. In 1976 the Saskatchewan Geological Survey re-examined the area and, although they came to the same conclusion regarding economic potential, Harper [89] discovered that a substantial drop in the Lake Athabasca water level over the preceding decade offered an unprecedented opportunity to determine the stratigraphy of the area, which is shown in Figure 2.6.

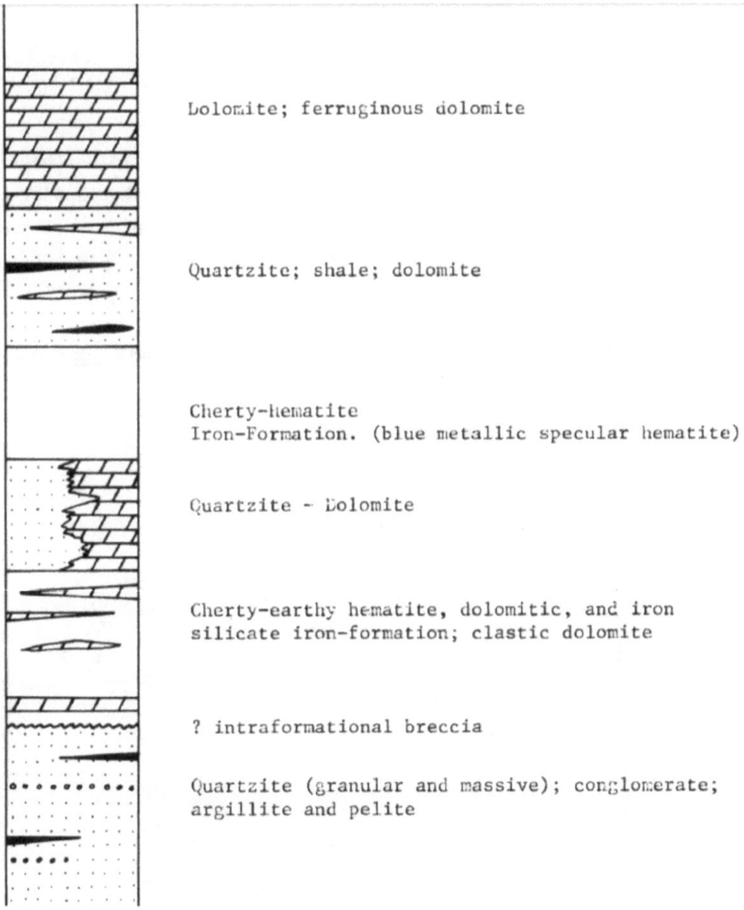

	Dolomite; ferruginous dolomite
	Quartzite; shale; dolomite
	Cherty-hematite Iron-Formation. (blue metallic specular hematite)
	Quartzite - Dolomite
	Cherty-earthy hematite, dolomitic, and iron silicate iron-formation; clastic dolomite
	? intraformational breccia
	Quartzite (granular and massive); conglomerate; argillite and pelite

Figure 2.6. Illustration of the stratigraphy of the Fish Hook Bay region. (Saskatchewan Geological Survey, 1976 [89].)

2.2 Prospecting for Base Metals.

The "Nicholson" area near the small bay between Cornwall Bay and Fish Hook Bay was re-staked in 1929, this time by the Mineral Belt Locators Syndicate. John D. Nicholson, who had just been made manager of Mineral Belt Locators Syndicate, sent the staking crew there in 1929 and Nicholson and his crew began development work at this location in 1930, looking for base metals. [90,91]. They identified a copper deposit at the site in 1930 [76,88], and this location subsequently became known as the "Nicholson Copper Property" [47] – see Figure 2.7. The copper minerals present in the Nicholson Copper Property are described by Robinson [76].

Figure 2.7. Rock sample, with copper showing, from the Nicholson Mine. (Author photo, 2017. The long dimension is 10 cm.)

Although niccolite (a nickel arsenide mineral, NiAs, see Figure 2.8), and pitchblende (a uranium oxide mineral, principally UO_2,) were also discovered at the site between 1929 and 1930, this timing coincided with the North American slump in the capital markets. As a result, Nicholson was unable to raise enough money to properly develop the property, although he did barge-in additional supplies and continued to work on the property each year, until 1934 [90].

Figure 2.8. Rock sample, with niccolite showing, from the Nicholson Mine. (Author photo, 2017. The long dimension is 12 cm.)

In 1933, gold was discovered in nearby Lodge Bay by Gus Nieman and Tom Box (for whom the Box Mine was later named). The ore was found to contain not only gold, but nickel, copper, molybdenum, lead and silver, prompting a mineral exploration rush in the area and the settlement (in 1937) of the town of Goldfields[21] near the Box Mine [1,43]. Upon learning of the Goldfields discovery and activities, Nicholson re-established his old claims in the fall of 1933 and prospected the site yet again, this time searching for gold [17,81,83,85]. Lang reports that local prospectors pointed out the sites pitchblende showings to him [87], and Nicholson himself is quoted as recollecting that he found *"Free gold, as well as more pitchblende and niccolite"* on his claims [90]. In 1935 he decided to create a company to develop the old Mineral Belt Locators claims, and J.D. Nicholson Mines Ltd. was incorporated in November of 1936[22] [78]. It was renamed

[21] The hoped-for gold bonanza did not materialize, but the name "Goldfields" stuck [10].

[22] The mineral claims on the site seem to have been transferred to the Nicholson company in 1937 [83].

Nicholson Mines Ltd. in January, 1937.

Preparing to develop the site for trenching and diamond drilling was not a simple matter. Being in a remote location, and without road access, most construction and other materials had to be brought in by boat, barge, or by air[23] (with aircraft landing on the lake). In 1931 Nicholson purchased a twenty-five foot fishing boat, and constructed his own freight barge. In his biography he relates that due to the high costs of goods and materials, he was able to provide boat and freight delivery services for others, whether for cash (for supplies like aviation fuel) or for trade (such as in exchange for "… 'hot' tips on new discoveries … [from] various prospectors") [90]. There were so many people wanting to get into the region to join in the prospecting wave that Nicholson was also able to get barge crews that would work their passage "in exchange for transportation of themselves and their baggage" [90].

The preferred, and most economical, option was to use a barge service. Barges were towed by shallow-draft tug boats, such as those of the Hudson's Bay Co. or the Northern Transportation Co. [92,93]. The principal barge route at the time was about 440 km (265 miles) along Lake Athabasca to the Athabasca River and then to the railhead at Waterways, Alberta, which in turn was served by the Northern Alberta Railway [8,65,94] (Figures 2.9 and 2.10). This provided about six months of ice-free waterway per year, but only about four months per year were practical for waterborne transportation in light of high winds and low water levels each fall season [31,66,95]. For example, according to Quiring [96], "a boat towing eight barges loaded with material for Gunner and Eldorado froze into the ice in 1956."

Eldorado found that the barging conditions ranged from "shallow bar ridden rivers to deep and frequently rough lakes," and that late summer water levels could fall as low as twenty inches in some areas, restricting the weight of loads that could be carried, and therefore increasing the cost per tonne shipped [97]. Despite these obstacles, transportation by barge continued to be the preferred option for the Nicholson operation throughout construction and the operating lifetime of the mine and campsite.

The Radium Line. Northern Transportation Co. operated a series of barge-towing tugboats, called "The Radium Line," that operated along the Mackenzie, Slave, and Athabasca Rivers from Tuktoyaktuk in the north, to Waterways in the south [98]. Along the way, barges could traverse Great Bear Lake to the uranium mine at Port Radium, Great Slave Lake to the gold mine at Yellowknife, and Lake Athabasca to the uranium mines in the Uranium City and Beaverlodge areas (Figure 2.9).

---

[23] Operated mostly by McMurray Air Services, which had a seaplane base at Uranium City.

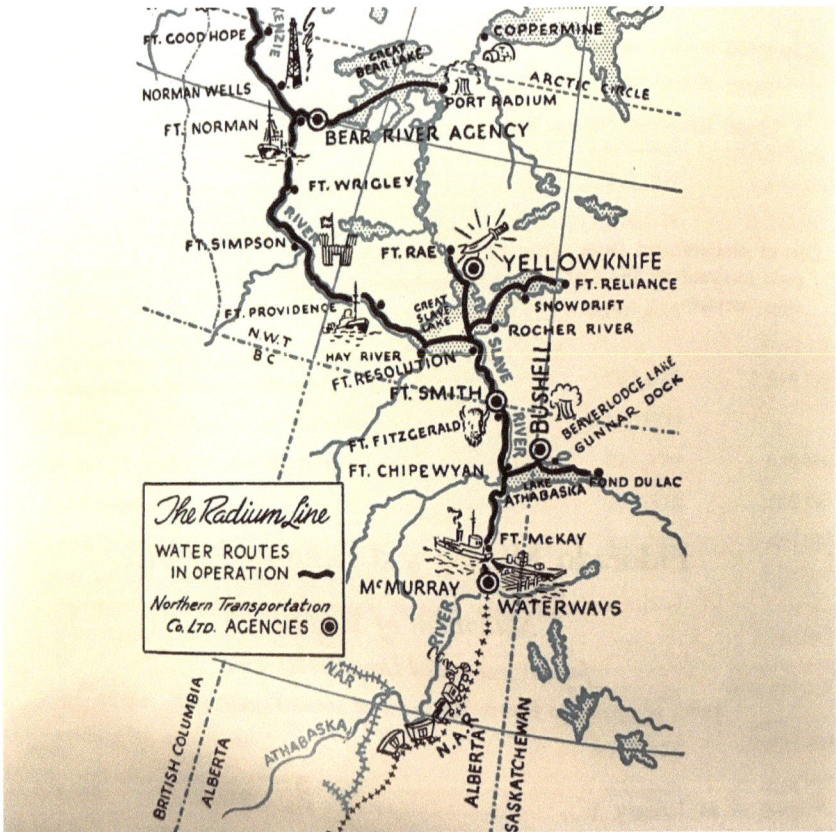

Figure 2.9. Illustration of barge routes northward from the railhead at Waterways, Alberta. Not shown is the leg from Fort Norman to Tuktoyaktuk. Adapted from a drawing in reference [98].

Figure 2.10. The Hudson's Bay Co. boat SS Distributor with a barge, carrying freight between Waterways, Alberta and Goldfields, Saskatchewan in 1937. Library and Archives Canada (Photo 3399965).

Even later, in the 1950s, the cost of freight by air was more than double that by water. As a result, mining companies in this region would attempt to order a year's worth of supplies prior to the beginning of the boat/barge season[24] for delivery in advance of winter freeze-up [1]. As a "rule of thumb" it was found that both capital and operating costs for the Athabasca region mines and mills, including Nicholson, tended to be more than double those of comparable southern operations [1].

During the 1935 search for gold, a zone (now known as Zone 4) was explored by digging two adits with approximately 107 m of underground workings, with discouraging results [81,85]. However, in July of 1935 during the course of this work, uranium secondary mineral staining and pitchblende were discovered in two locations on the Nicholson site, one of them being on the JIM claim [14]. Nicholson and his prospectors recognized pitchblende at the site and this was picked up almost immediately in media reports (Figure 2.11, reference [99]).

---

[24] Ice breakup would usually occur in late May, with some areas not clearing until mid-June, and the lakes would usually be frozen-over by the end of October yielding an effective barge season of about 15 weeks each year [15,31].

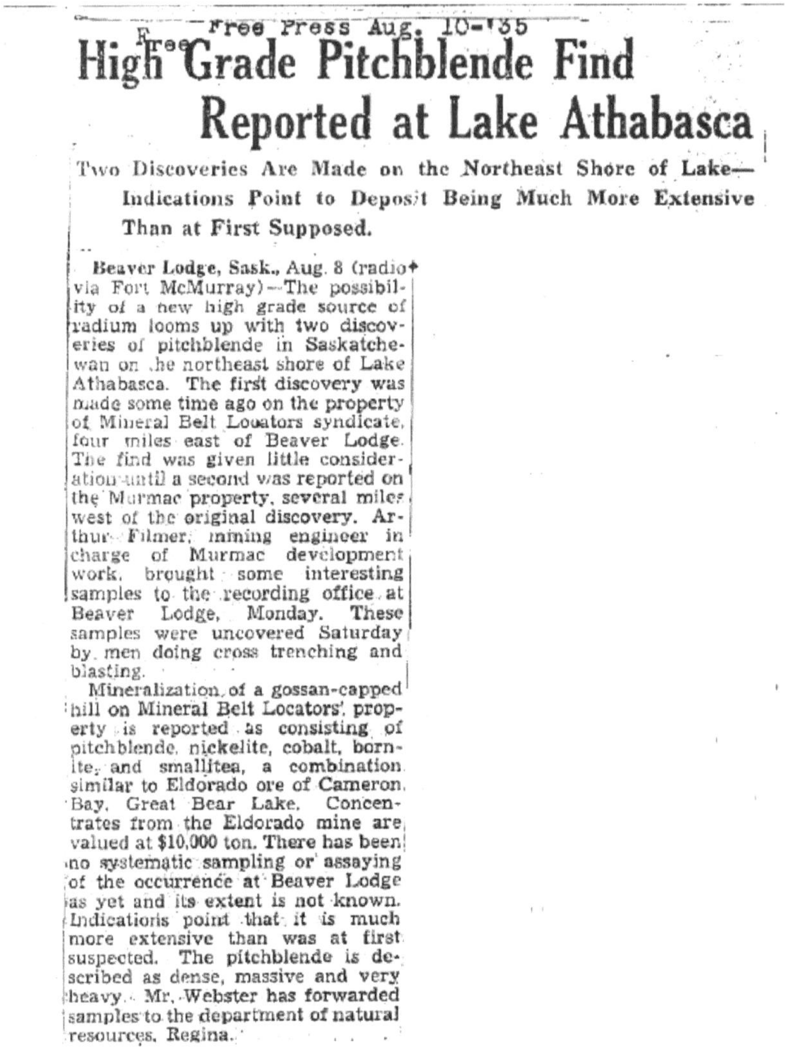

Free Press Aug. 10-'35

High Grade Pitchblende Find
Reported at Lake Athabasca

Two Discoveries Are Made on the Northeast Shore of Lake—
Indications Point to Deposit Being Much More Extensive
Than at First Supposed.

Beaver Lodge, Sask., Aug. 8 (radio
via Fort McMurray)—The possibility of a new high grade source of
radium looms up with two discoveries of pitchblende in Saskatchewan on the northeast shore of Lake
Athabasca. The first discovery was
made some time ago on the property
of Mineral Belt Locators syndicate,
four miles east of Beaver Lodge.
The find was given little consideration until a second was reported on
the Murmac property, several miles
west of the original discovery. Ar-
thur Filmer, mining engineer in
charge of Murmac development
work, brought some interesting
samples to the recording office at
Beaver Lodge, Monday. These
samples were uncovered Saturday
by men doing cross trenching and
blasting.

Mineralization of a gossan-capped
hill on Mineral Belt Locators' property is reported as consisting of
pitchblende, nickelite, cobalt, born-
ite, and smallitea, a combination
similar to Eldorado ore of Cameron
Bay, Great Bear Lake. Concentrates from the Eldorado mine are
valued at $10,000 ton. There has been
no systematic sampling or assaying
of the occurrence at Beaver Lodge
as yet and its extent is not known.
Indications point that it is much
more extensive than was at first
suspected. The pitchblende is described as dense, massive and very
heavy. Mr. Webster has forwarded
samples to the department of natural
resources, Regina.

Figure 2.11. Free Press news headline in 1935 reporting the pitchblende find at the Mineral Belt Locators' Nicholson site [99].

Nicholson also brought the showings to the attention of Alcock, whose 1936 Geological Survey of Canada report would later bring other interests to the area [14,83]. In his report, Alcock noted that *"Free gold occurs in the quartz of the veins and stringers and in places visible gold has been found in the granite itself away from the vein quartz"* [14]. This is the origin of the name "Goldfields" for the area. Nicholson's work developing the site for gold continued for several years (Figure 2.12).

NICHOLSON MINES LIMITED
1331 CANADIAN BANK OF COMMERCE BUILDING
TORONTO, CANADA

July 20, 1939.

Department of Mines and Resources,
Ottawa,
Ontario.

Gentlemen:-   Attention: Mr. R. A. Gibson
                       File 16151 TML
                            14986 TML
                            14987 TML

         Replying to yours of June 26th in
reference to excess acreage on the two gold claims
Nos. 2 and 3, we are enclosing our cheque in the
amount of $235.00 to cover this excess penalty.

              Yours very truly,

              NICHOLSON MINES LIMITED

              L. A. Macdonald

Encl.   ass't Secretary-Treasurer.
LAM/C.

Figure 2.12. Example of Nicholson Mines correspondence regarding their gold exploration activities at the Nicholson mine in the 1930s. (Library and Archives Canada)

By 1941, Nicholson Mines had decided that further work was "not warranted" at that time but renewed geophysical examination might be pursued at a later date [100]. Things declined at nearby Goldfields as well in that the Box mine had played-out by 1942 and both it and the town of Goldfields itself were abandoned in 1942 [8]. Activity would return to all of these areas in later years. In the meantime, Nicholson Mines focused their attention on properties in the Northwest Territories [100].

A/Supt. John D. Nicholson

The year 1940 may have marked the final year of John Nicholson's direct personal involvement with the Nicholson Mines, at which time he had just moved on to yet another chapter of an incredible career, of which only a very brief summary is provided here.

John D. Nicholson was born in Provincetown, Massachusetts, in 1863, and moved to Canada as a boy. He initially worked on a series of ships, travelling first along the Canadian east coast, then to South America, and later to Hudson's Bay. He joined the North West Mounted Police (NWMP) in 1885, serving mostly in Saskatchewan and Alberta. In 1899, he was granted "leave of absence without pay" in order to join the Lord Strathcona's Horse & 5th Canadian Mounted Rifles and serve overseas in what at the time was known as the South African War[25]. After the war, Nicholson resumed his work with what was then named the Royal North West Mounted Police (RNWMP, see Appendix 4), where he volunteered for northern duty and served until his first retirement, in 1911. In 1907, he received a commendation for his work in the apprehension of cattle stealers in Alberta.

The Nicholson Mine's Namesake: A/Supt. John D. Nicholson

1885 – 1899	North West Mounted Police (NWMP),
1899 – 1902	5th Canadian Mounted Rifles, South African War,
1902 – 1911	Royal North West Mounted Police (RNWMP),
1911 – 1917	First 'retirement,'
1917 – 1927	Alberta Provincial Police (APP),
1927 - 1929	Second 'retirement,' prospecting,
1929 - 1936	Field Manager, Mineral Belt Locators,
	Developed the "Nicholson Copper" property,
1937 - 1939	Joined the NWT Gold Rush,
1939 – 1942	Royal Canadian Mounted Police (RCMP),
	Security Intelligence Service,
1942 - 1945	Third retirement.

When the Alberta Provincial Police (APP) force was created in 1917, Nicholson joined and was made Assistant Superintendent (Figure 2.13). He was also appointed Chief Detective for the APP, and organized its Detective Division. He served in the APP until his second retirement in 1927.

---

[25] Now more commonly known as the Second Boer War, 1899 - 1902.

Once again needing something to do in "retirement," Nicholson turned to prospecting in northern Manitoba, northern Saskatchewan, and the Northwest Territories. Along the way, he discovered new mineral deposits and became engaged in barging supplies on the Athabasca, Slave, and Mackenzie Rivers, as well as on Lake Athabasca. Invited to join the Mineral Belt Locators Syndicate as field manager in 1929, as noted above, he quickly focused on the area near the small bay that would eventually bear his name, located between Cornwall Bay and Fish Hook Bay. In 1930, they built the first cabins on the site and spent the summer "blasting into the showings" [90]. In 1934 Nicholson participated in a "Prospectors' School," and founded the North West Prospectors Association, becoming its first President [90]. He continued to work on the "Nicholson Copper" property until 1935. In 1936 they continued to find free gold, uranium, and nickel.

Figure 2.13. Assistant Superintendent John D. Nicholson (Alberta Provincial Police) is shown seated to the left in this photograph circa. 1925. Also shown is Superintendent Bryan (centre) and Inspector Bavin (right). (Provincial Archives of Alberta, Photo A4821.)

In 1937 he joined the NWT Gold Rush to search for gold in the Yellowknife area, continuing to work there until 1939, when Germany invaded Poland, effectively beginning the Second World War. With the outbreak of war in 1939, the Royal Canadian Mounted Police (RCMP, see Appendix 4) found themselves short-handed and increased their recruiting efforts. Although Nicholson was by then seventy-six years old, he offered his services to the RCMP and was accepted. With this transition, he resigned from his position as President of Nicholson Mines in May of 1940 [101]. He served in the RCMP security intelligence service, doing anti-espionage work, until 1942, when he retired for the third and last time, on the eve of his eightieth birthday. He received the Queen's and King's medals, as well as the RCMP Long-Service Medal. He died in Victoria, in 1945. Additional details can be found in references [90,102-106], particularly in a biography written by J.W. Horan (Figure 2.14; reference [90]).

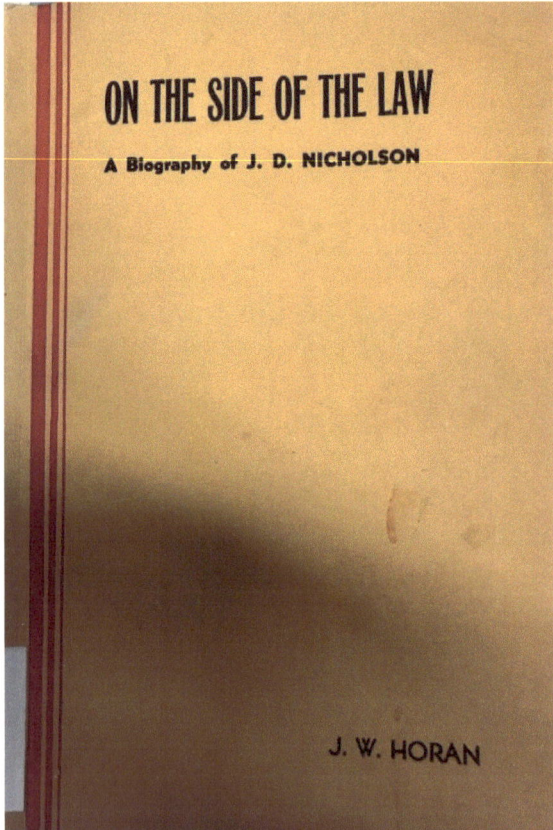

Figure 2.14. A Biography of J.D. Nicholson, 1944 [90].

2.3 Prospecting for Gold and Uranium.

As noted above, when the Nicholson site was being explored for gold in 1935, uranium stain and pitchblende were discovered while digging [85]. This represents the first discovery of uranium in the Lake Athabasca region, and the first in Saskatchewan [16].

As described in Section 1.3, the 1940s had brought another surge in Canadian uranium exploration, especially once private enterprises were again allowed to get involved in uranium exploration and mining. In addition, technology advances had helped to make low-grade (~0.1 %) ore deposits economically viable. In the early 1940s, the only known Canadian pitchblende occurrence apart from the Great Bear Lake region in the Northwest Territories, was the one at the Nicholson site[26] [83]. For this reason, the newly established strategic importance of uranium to Canada in 1944, led A.W. (Fred) Jolliffe, of the Geological Survey of Canada, and R. (Dick) Murphy, of Eldorado Mining and Refining Ltd. to re-examine the Nicholson site, as a result of which Eldorado launched an intensive prospecting program in the area [10,83,87,88,107] as did Nicholson Mines itself, in 1944 and 1945 [108,109].

For its part, Eldorado staked 22 claims east and northeast of the Nicholson property in 1944 and resolved to launch a systematic prospecting effort over the next two years [87,107]. For this purpose, Eldorado established a camp at Fish Hook Bay, "where pontoon planes could land" [10]. Where radioactivity was encountered, the more promising areas were stripped and trenched, and then explored further with diamond drilling [83]. Altogether, hundreds of pitchblende occurrences were found in the Beaverlodge region [1,107]. As a result, Eldorado's Fish Hook Bay camp was abandoned in the winter of 1948, and the buildings were moved over the ice to Eldorado's more permanent Beaverlodge mine and mill site, just outside of Uranium City [87].

In 1948 the federal government repealed the ban on prospecting for uranium by the general public and in 1949 the Saskatchewan government established and sold prospecting concessions in the Beaverlodge area[27] [88]. This led to both government and private prospectors to scour the region, including at the Nicholson site, where Nicholson Mines resumed prospecting work in 1948 [83,87,110]. Eldorado's official history [1] notes that:

"Between 1948 and 1953 swarms of prospectors roved far and wide in Canada, many of them amateurs with the exaggerated idea that a few clicks of a Geiger

---

[26] At this point in time, the occurrence in Ontario, near Lake Superior, was still unconfirmed.

[27] These were mostly 65 km^2 (25 square mile) areas, sold for about $50,000 each [83].

counter would make them rich. There is no record of a really worthwhile discovery having been made by such people, but a number of professional and truly knowledgeable prospectors did 'strike it rich.'"

In 1948, additional prospecting at Nicholson site led to cleaning and re-examination of the old trenches in zone No. 4 along with the digging of new trenches, exposing a further 335 m of the zone (1,100 feet) [78,111]. Nicholson Mines also conducted more diamond drilling with Geiger probing of the kind illustrated in Figure 2.15 [110,111]. A typical core sample is shown in Figure 2.16.

Figure 2.15. Geologists "Geiger probing" diamond drill holes. This photo is from La Ronge Uranium Mines, at Nistowiak", Saskatchewan, in 1952 but a similar scene at the Nicholson Mines would have looked much the same. Glenbow Museum Archives. (Photo PA-2218-540).

According to Lang, a test shipment of *"sorted material from the trenches and the upper adit was sent to the Eldorado refinery, and payment was made for high contents of uranium, gold, and silver"* [78] (the assay work was conducted at Eldorado's refinery at Port Hope, Ontario). Other samples were sent to the Bureau of Mines in Ottawa, presumably for mineralogical analyses. The 1948 season's work yielded promising results for gold and encouraging results for uranium (Figure 2.17) [112-114]. Encouraged by these results, Nicholson Mines started shipping "shaft-sinking equipment and supplies" over the frozen Lake Athabasca[28] during the winter of 1948/49 [110].

---

[28] The official spelling of the name "Lake Athabaska" was changed to "Lake Athabasca" in 1948 [78,87].

Figure 2.16. Diamond-drill, "one inch" core sample from the Nicholson Mine. The spike is from the underground mine railway described in Chapter 3. (Author photo, 2017. The core diameter is 3 cm.)

Between 1948 and 1949, additional blasting, trenching, and 15 drill holes were completed in zone No. 4, combined with another 26 drill holes among the other zones [78,81,85] (see Figure 2.17). Blasting (Figures 2.18 and 2.19) was principally carried out with C.I.L. 70 % Forcite/Driftite in 1" and 7/8" diameters [115]. This was a "gelatin dynamite," comprising 70% nitroglycerin mixed with cellulose, sodium or potassium nitrate, and a hydrocarbon like tar (to make it waterproof).

Encouraging Report From Nicholson

J. D. Mason, who is devoting his full effort this fall and winter to the development of the Goldfields property of Nicholson Mines Ltd., was in Yellowknife for a few days this week, with the company's geologist, Dr. G. G. McCartney of Toronto and diamond drill contractor J. C. Kelly. The report from this property, which carries both gold and uranium, continues to be of much interest.

Mr. Mason plans to continue work there until early in the new year, and expects to have a heavy program also for 1949.

Two zones, Nos. 2 and 4, 2,200 feet apart, have been under exploration. No. 4 zone has been exposed by trenching and stripping over a distance of 500 feet and in mid-September a 175-foot-long section was being sampled. Assays are said to indicate 1.71 per cent uranium and 0.15 ounce gold across three feet mean width. No. 2 zone has been

NICHOLSON ASSAYS SHOW HIGH GOLD

Returns Received Lengthen the Oreshoot in No. 2 Zone and Improve Grade

Assay returns received by Nicholson Mines Ltd. on approximately 50 of the 120 samples sent to the Department of Mines at Ottawa for assay have more than lived up to expectations, The Northern Miner learns. The gold values are particularly good with a number of the assays running in the ounces. Uranium assays have confirmed the previous sampling but on the whole might run a little bit lower. No assays have yet been received on the platinum-palladium content.

Most of the assays received so far are for samples taken on the No. 2 zone and preliminary calculations now indicate a length of 180 ft. averaging 2.75 ozs. gold, cut to 1.28 oz., and 0.56% uranium oxide

(Continued on Page Sixteen)

NICHOLSON REPORTS URANIUM IN ADIT

Examination of Old Work Reveals Seam Carrying Pitchblende—More Assays Soon

Nicholson Mines Ltd. reports that just prior to freeze-up the lower tunnel at a depth of 50 ft. on the No. 4 showing was reopened. Drifting at this level followed the vein zone for a length of 105 ft. The main pitchblende-bearing seam showing fresh unoxidized pitchblende, is exposed at both ends of the drift and is continuing strong in both faces. The two end sections exposing the pitchblende seam represent about half the total length of drifting. Along the central section of the drift, the main seam is in the hanging wall.

The drift was driven on former operations, prior to war, in order to check on gold values if the calcite filled breccia zone and the main pitchblende seam was not followed for this reason. However, there is evidence that the main seam extends for the entire drift length. Two parallel seams show in the central section of the drift but it has not been definitely determined whether these carry uranium. Geiger counter reactions are not reliable owing to the proximity of the main seam. Some samples from this drift are included in those row at Ottawa for analysis.

A long diamond drill hole is currently being drilled. Progress is slow as the hole is being cased for its entire length in an endeavour to obtain reliable sludge samples. At last report (November 5th) the hole was at a depth of 400 ft. with 300-350 ft. to go to its objective.

Expect Assays Next Week

The shipment of hand sorted and cobbed uranium ore has arrived at the Eldorado refinery, Port Hope. This shipment consists of five lots with a total weight in excess of 3,000 lbs. The lot will be assayed separately and are in the nature of test shipments for determining metallurgical data. They were blasted from surface pits on the No. 2 and No. 4 zones. A such test lot, from No. 2 Zone, is en route from Waterways, Alberta, to the Port Hope refinery. Assays for the uranium, gold and silver content of the shipment should be available in eight to 10 days. Platinum-palladium determinations will take one month.

Over 100 samples, representing sampling of additional lengths on the No. 2 and No. 4 zone, are at the Bureau of Mines, in Ottawa, for analysis. Results on these should be available in about one week's time. An enormous amount of chemical work is involved in these analyses. The company advises that Department of Mines is extremely co-operative in expediting this work as greatly as possible.

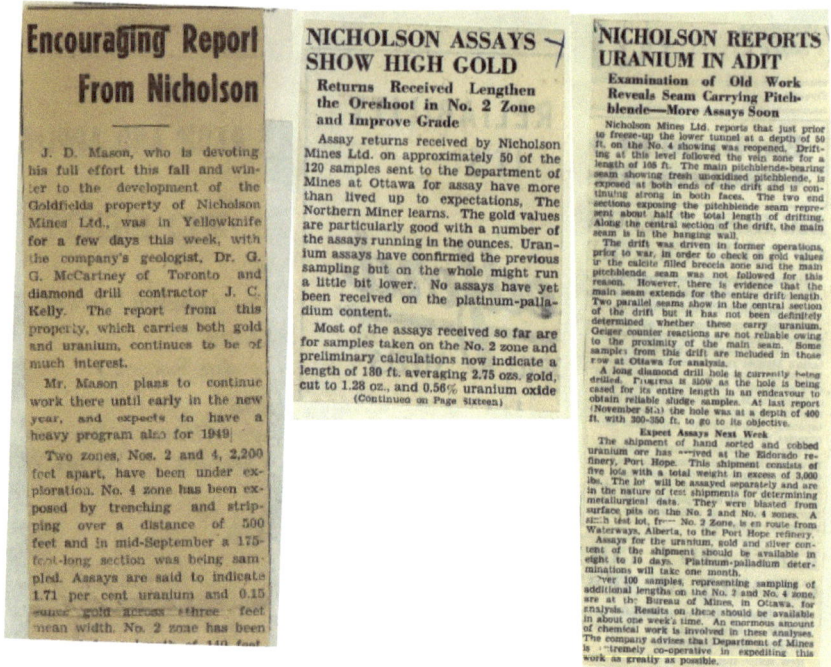

Figure 2.17. Examples of *News of the North (Yellowknife)* and *Northern Miner* news headlines in 1948 reporting the early Nicholson site exploration and development results for gold and uranium [112-114].

It was ultimately decided to develop two mine shafts[29], on each of zones No. 2 and 4, which was carried out between 1949 and 1950 [76] (Figures 2.18 through 2.20). These developments indicated that there were "rich shoots" in the zone No. 4 area [83] and Nicholson reported receiving encouraging results for both gold and uranium, but by this time the major focus was on uranium [116,117] (see Figure 2.20).

A mine shaft with two lateral developments (at the 30 and 60 m levels) was sunk to 71 m in zone No. 4 [81]. By the end of 1950, approximately 466 m of lateral work had been completed at the first (30 m) level, and approximately 408 m at the second (60 m) level. Similarly, a 41 m mine shaft with lateral development at the 30 m level was completed in zone No. 2, and an 18 m shaft without further development was completed in the zone No. 1 extension area [81].

---

[29] The Atomic Energy Control Board would have granted a mining permit in 1949 authorizing the development and mining.

Figure 2.18. Miner Jim Mills, shown here holding dynamite sticks in September 1949, had come from Timmins, Ontario to work at the Nicholson Mine. Glenbow Museum Archives. (Photo NA-4510-24).

As the results of the zone Nos. 1 and 2 developments were not considered encouraging, they were shut down. The focus then shifted to the more promising zone No. 4 mine [17], for which very positive results were reported [119] – so much so that consideration was given to the feasibility of constructing a dedicated uranium mill on the Nicholson site [118] (Figure 2.20).

In May of the next year, 1951, Nicholson Mines Ltd. was reorganized and incorporated as Consolidated Nicholson Mines Ltd.

Figure 2.19. A surviving case that would have held sticks of Forcite explosive. (Saskatchewan Research Council, 2015.)

By 1952, despite numerous waves of drilling and exploratory blasting in the shafts, the company had not yet discovered enough ore to justify building a mill onsite, however plans were in place to deepen the main shaft and explore on two additional levels "with the hope of establishing enough ore for that purpose" [78]. Although plans for a mill did not ever materialize, significant mining activities were carried out at both levels of the zone No. 4 mine[30]. In late summer of 1954 the company obtained a contract for the sale of ore to Eldorado Mining and Refining Ltd. [81], and the first shipments of uranium ore to Eldorado occurred in 1954 [120,121].

---

[30] Additional drilling was conducted at zone No 1 during 1955, but once again the results were not found to be encouraging.

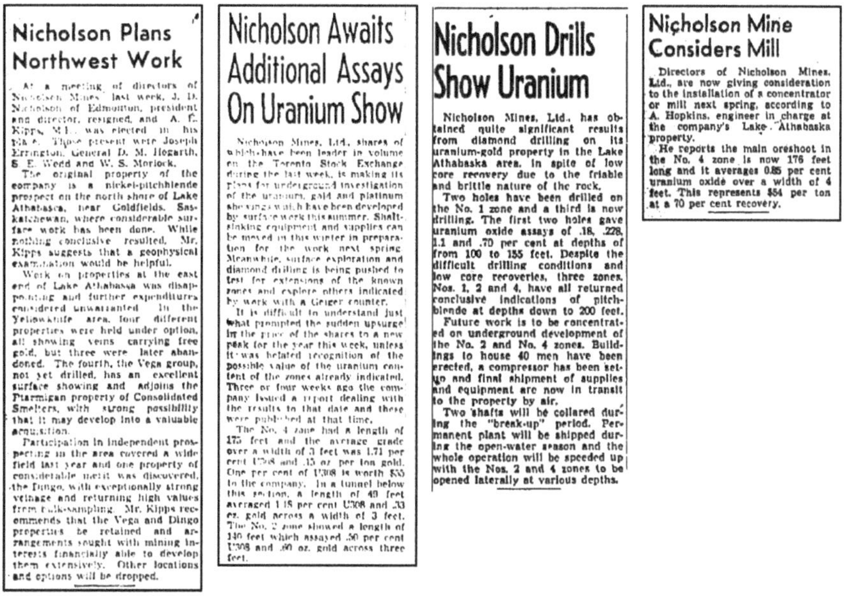

Nicholson Plans Northwest Work

At a meeting of directors of Nicholson Mines last week, J. D. Nicholson of Edmonton, president and director, resigned, and A. F. Kipps, M.I. was elected in his place. Those present were Joseph Errington, General D. M. Hogarth, S. E. Wood and W. S. Morlock.

The original property of the company is a nickel-pitchblende prospect on the north shore of Lake Athabasca, near Goldfields, Saskatchewan, where considerable surface work has been done. While nothing conclusive resulted, Mr. Kipps suggests that a geophysical examination would be helpful.

Work on properties at the east end of Lake Athabasca was disappointing and further expenditures considered unwarranted. In the Yellowknife area, four different properties were held under option, all showing veins carrying free gold, but three were later abandoned. The fourth, the Vega group, not yet drilled, has an excellent surface showing and adjoins the Ptarmigan property of Consolidated Smelters, with strong possibility that it may develop into a valuable acquisition.

Participation in independent prospecting in the area covered a wide field last year and one property of considerable merit was discovered, the Dingo, with exceptionally strong values from bulk-sampling. Mr. Kipps recommends that the Vega and Dingo properties be retained and arrangements sought with mining interests financially able to develop them extensively. Other locations and options will be dropped.

Nicholson Awaits Additional Assays On Uranium Show

Nicholson Mines, Ltd., shares of which have been leader in volume on the Toronto Stock Exchange during the last week, is making its plans for underground investigation of the uranium, gold and platinum showings which have been developed by surface work this summer. Shaft-sinking equipment and supplies can be moved in this winter in preparation for the work next spring. Meanwhile, surface exploration and diamond drilling is being pushed to test for extensions of the known zones and explore others indicated by work with a Geiger counter.

It is difficult to understand just what prompted the sudden upsurge in the price of the shares to a new peak for the year this week, unless it was belated recognition of the possible value of the uranium content of the zones already indicated. Three or four weeks ago the company issued a report dealing with the results to that date and these were published at that time.

The No. 4 zone had a length of 175 feet and the average grade over a width of 3 feet was 1.71 per cent U308 and .15 oz per ton gold. One per cent of U308 is worth $55 to the company. In a tunnel below this section, a length of 40 feet averaged 1.18 per cent U308 and .33 oz. gold across a width of 3 feet. The No. 2 zone showed a length of 140 feet which assayed .50 per cent U308 and .90 oz. gold across three feet.

Nicholson Drills Show Uranium

Nicholson Mines, Ltd. has obtained quite significant results from diamond drilling on its uranium-gold property in the Lake Athabaska area. In spite of low core recovery due to the friable and brittle nature of the rock.

Two holes have been drilled on the No. 1 zone and a third is now drilling. The first two holes gave uranium oxide assays of .18, .228, 1.1 and .70 per cent at depths of from 100 to 155 feet. Despite the difficult drilling conditions and low core recoveries, three zones, Nos. 1, 2 and 4, have all returned conclusive indications of pitchblende at depths down to 200 feet.

Future work is to be concentrated on underground development of the No. 2 and No. 4 zones. Buildings to house 40 men have been erected, a compressor has been set up and final shipment of supplies and equipment are now in transit to the property by air.

Two shafts will be collared during the "break-up" period. Permanent plant will be shipped during the open-water season and the whole operation will be speeded up with the Nos. 2 and 4 zones to be opened laterally at various depths.

Nicholson Mine Considers Mill

Directors of Nicholson Mines, Ltd., are now giving consideration to the installation of a concentrator or mill next spring, according to A. Hopkins, engineer in charge at the company's Lake Athabaska property.

He reports the main oreshoot in the No. 4 zone is now 176 feet long and it averages 0.85 per cent uranium oxide over a width of 4 feet. This represents $54 per ton at a 70 per cent recovery.

Figure 2.20. Illustration of Canadian news headlines (1940-1950) reporting the early Nicholson site uranium exploration and developments [91,99,104,118].

Between 1952 and 1954 the sense of a booming industry in Northern Saskatchewan, combined with announcements regarding the Beaverlodge area uranium mines in particular, again caught world attention with media coverage from North America [122-126], to as far away as Australia [127,128] (see Figure 2.21). This was even captured in broadly circulated magazines like Maclean's and Life [49,129,130].

In late summer of 1954 the company obtained a contract for the sale of ore to Eldorado Mining and Refinement Ltd., and the mine was reopened in 1955 [81]. Also in 1955, Nicholson Mines leased another 54 adjoining claims to the north and east from Eldorado, comprising Eldorado's Fish Hook Bay property [131].

Canada looks to active uranium year

SOME indication of the work ahead of Australian uranium prospecting companies is provided by a cabled summary of progress and prospects in Canada received yesterday from Uranium City.

The United Press reports that activity in Canada's richest proven uranium field mushroomed in 1953, and that authorities predict "limitless" possibilities for 1954.

The federally owned and operated Eldorado Mining and Refining Ltd. began producing the atomic fuel at Uranium City, Saskatchewan, on June 1.

Private companies then spurred development and a few of them started mining operations before the end of the year.

Total expenditures by these private companies topped the 10 million dollars mark.

Gunnar Gold Mines held the mineral spotlight in the north for most of the year. It staked close on 100 million dollars worth of mining ore, and now is preparing a low-cost, open-pit mining operation.

The company plans a seven million dollars capital expenditure to bring it to the mining stage, possibly late in 1954.

Gunnar also will operate a second concentrator plant in the north with twice the milling capacity of the present Eldorado plant.

More than 11,000 claims were staked during the year. Many of the private companies could be producing uranium today, except that Eldorado, which has the only concentrator plant in the area, will be unable to handle the ore until early in

handle the ore until early in 1954

Saskatchewan Resources Minister, Mr. J. H. Brockelbank, put it this way: "Although 1953 was a big year—1954 promises to be an even bigger one."

Canada has a "Rum Jungle"

from a Special Correspondent

URANIUM CITY (Lake Athabaska) — A French-Canadian prospector called Monsey Mercredi tried to sell me a part share in a uranium claim for a crate of bottled beer. The claim was genuine enough.

It lies out beyond Martin Lake, where radio activity makes red rocks crackle like fried eggs. But beer is usually worth more than uranium among those who peddle property in this shack town of atomic dreams.

Beer can be drunk. It also costs up to a dollar a bottle "in the bush" by contract. Uranium is everywhere. And almost everyone in Canada's Uranium City (population 1150) has got some.

No mining

WITHIN two hours of my arrival on the old Athabaska steamer (it makes the trip twice a year) I heard about good property found by Father O'Brien, the young priest here, who has the reputation of being the best husky dog team "musher" in the North.

I also inspected a solid hunk of black uranium ore about the size of a bowler hat, produced by Gus Hawker, the local storekeeper and fabulous character.

Ted Ellingham, Government geologist, told me that about 12,000 claims have been formally registered in the last 18 months. The snag is that nobody seems to be mining the stuff.

Despite rumours of rivalry and rumpuses in this bewildering boom-town, there are only three mines producing uranium in the whole of Canada.

Those mines are at Port Radium, on the Great Bear Lake, at the huge Government-controlled mine of El-

Lake, at the huge Government-controlled mine of Eldorado at Beaverlodge, just over the hill, and at the Rix Athabaska holdings, a few miles away.

Rumours rife

SOME mines may hit the atomic jackpot in the future. Gunnar, for instance. It lies on the edge of Lake Athabaska.

Gilbert Labine, of Ottawa owns it, and has just spent 500,000 dollars on trucks and shovels in a preliminary ground-skinning operation.

Other mines are a constant source of rumour in Uranium City's huge saloon bar.

Jim Lang, the Mountie chief, rarely arrests any drunk who is not actually wrecking the town. There are too many of them.

Uranium City is a man's town. Women, including copper-skinned Indian girls, are outnumbered by four to one. It is also a shack town, despite one fine hotel school, post office, and church.

Figure 2.21. Illustration of 1950s Australian news headlines[31] reporting the Saskatchewan uranium mine development boom [127,128].

---

[31] The Rum Jungle referenced in the right-hand article was a uranium deposit in the Northern Territory, Australia. Discovered in 1949, a mine and mill were constructed in 1952 that operated from 1953 to 1971.

Stope mining (see Chapter 3) was conducted at both levels of the zone No. 4 mine[32]. All the ore mined in 1955 was stockpiled to wait for the lake to freeze, although surveying had also begun with the aim of eventually building a road to the Eldorado mill at Beaverlodge[33][131]. During the winter of 1955/56 about 1,800 tonnes (2,000 tons) of mined uranium ore, averaging 0.304% U_3O_8, was trucked 24 km (15 miles) across Lake Athabasca, to the Eldorado mill. This was estimated to have a value of $551,817 [83]. By the time production had halted in 1956, the Consolidated Nicholson Mine had produced an estimated 11,000 tonnes (12,000 tons) of ore at an overall average grade of 0.304 % U_3O_8 [17].

Eldorado reports receiving significant shipments from Consolidated Nicholson in each of 1954, 1955, and 1956 [121,132-133]. Following the 1956 shipments of ore to Eldorado the deposit was not yet considered exhausted, however the contract with Eldorado had expired and Eldorado's offer of extension was apparently for too low a price for Nicholson to accept [83].

In 1957, with the discovery of a new vanadium mineral[34] on the Nicholson site, additional mineralogical work was done to explore the idea that there might be enough vanadium in the vein for economic development [134]. However, the economic assessment did not support further development for this metal.

This setback produced another lull in production, lasting until 1958 when Buddy Leask, Ed Kull, and Ted Kennedy obtained a lease on the Nicholson deposit, formed the KLK Mining Company, and attempted to high grade what was left (meaning that they went after only the highest-grade remnants that had been left behind) [135]. According to Kennedy [135]:

> *"The first thing we did was pick over the dump by hand, using a geiger counter, shovels and picks. We shipped one barge load of uranium ore that summer and did a little better than cover our costs. We stockpiled another bit, which we trucked over the ice the next winter."*

KLK also high graded the deposit above the lower adit level between 1958 and 1959, extracting about 1,800 tonnes (2,000 tons) of uranium ore, averaging 0.5% U_3O_8 [17,81,85,135]. As was the case in the 1940s, all of the picked and mined ore was taken to Eldorado for milling.

---

[32] Additional drilling was conducted at zone No 1 during 1955, but once again the results were not found to be encouraging.

[33] The road was never constructed, however.

[34] The new mineral, an iron vanadium oxide, was named "nolanite" in honor of geologist Dr. Thomas B. Nolan, Director of the U. S. Geological Survey.

In Kennedy's autobiography [135], he writes:

"We kept on drilling and blasting, we would tram the ore from underground in one ton cars, pull them up a ramp with a little air tugger and load the ore directly onto a 10 ton truck then during the winter months haul it over the ice of Lake Athabasca, over the Beaverlodge portage, over the ice of Beaverlodge Lake and to the Eldorado Mill at the north end of the lake. Once the ice was gone, we stockpiled the ore and when the stockpile was big enough we would order a barge from the Northern Transportation Company. Load up that barge as quick as we could because we were paying by the day as well as the ton, from the small stockpile and from freshly mined stuff. It held one hundred or so tons and we would call in the tugboat to be there at the time it had a full load. The tug would pull that barge around Crackingstone point, past Gunnar mines ... and on over to Bushell, the small port. Once there, a front end loader would pile that uranium ore onto a bunch of trucks and over to the mill at Eldorado, about 15 miles."

Kennedy also provides a semi-serious, generalized description of the business of building one of the small mines in the Uranium City area, in the mid-1950s uranium boom years [135]:

"A prospector goes out and stakes a claim. It's better if there is a showing of uranium but that part doesn't really matter too much so long as it shows on the claim map of Uranium City showing the Eldorado claims, sells it to a promoter who forms a company on it or puts it into an old company, gets a bunch of shares for himself and the prospector mostly for himself. Gets an engineer to write a report saying how good it is. Gets a lawyer to prepare and submit a prospectus which sets out in glowing terms the future of the company and the property. Gets a stock broker to underwrite the stock, and he sells the shares to his clients, gets listed on the Toronto stock exchange and the shares start trading. The promoter uses some of the money raised to get the geologist and prospector to do some work on the ground. This is easy to do because uranium can be detected with a Geiger counter or scintillometer and around that area 10,000 or so uranium showings were found so lots of finds were being made.

When a find is made, blast it with dynamite and get some good samples, assay them and publish the results, the geologist having written another glowing report. Everybody gets excited and buys stock, the price goes up and the promoter sells more stock to the public including most of his own and so raises more money. Now he might use that money to drill, if the results are good he can raise more dough and do more drilling and on it goes, if the results keep good then a mine will be established and everyone will make lots of money. For example Gunnar Mines. But usually it doesn't work out. Mining exploration is a business of failure and the vast majority of prospects end up on the rubbish heap."

Kennedy's final words were underscored by his own experiences mining the Nicholson deposit with the KLK Mining Company: *"That lasted a year until we went broke, EM&R[35] wouldn't take our ore any more"* [135]. The reason Eldorado stopped buying uranium ore was that, in 1959, the markets for uranium collapsed when the United States and United Kingdom stopped stockpiling uranium. Although the Canadian government continued to buy uranium to preserve some of the industrial activity, most of the uranium mines in Canada closed – Consolidated Nicholson and KLK Mining included.

In the 1960s and 70s, Enex Resources and Consolidated Nicholson Mines Ltd. completed additional exploration work in the same general area, including at zone No. 1, the No. 1 Extension, and the secondary zone No. 6. In 1964 control of Consolidated Nicholson was acquired by Lamaque Mining Co. Ltd., a subsidiary of Teck Corp. Ltd., although the company name was not changed [136]. On June 4, 1965, Consolidated Nicholson Mines Ltd. obtained a Quit Claim Deed for the surface lease. This separated the surface lease from the company, and it reverted back to the Crown.

Thereafter, the mineral ownership became a separate issue from the previous mine operations and infrastructure, and throughout the 1970s various companies tried additional exploration drilling around the site, including 8 new holes drilled for Imperial Oil again in 1975 [137], but without finding any economic mineralization.

In 1977, Consolidated Nicholson (which had by then become Auric Resources Ltd.) continued to explore between zones No. 1 and 2. In the 1980s, the company (which by then had become Chancellor Energy Resources Inc.) continued to explore throughout zones No. 1, 2, and 6, as did other companies including Denison and Eldor [85]. However, none of these exploration efforts proved successful.

What was once Nicholson Mines continued to go through multiple reverse splits and mergers and acquisitions over time, as illustrated in Table 2.2.

---

[35] Eldorado Mining and Refining Ltd.

Table 2.2. Nicholson Mines' Corporate Evolution [8,78,138,139].

Year	Corporate Change
1936	Incorporated as J.D. Nicholson Mines Ltd.
1937	Renamed Nicholson Mines Ltd.
1947	Nicholson Mines Ltd. becomes a subsidiary of Transcontinental Resources Ltd.
1951	Name changed to Consolidated Nicholson Mines Ltd.
1975	Name changed to Auric Resources Ltd.
1979	Name changed to Chancellor Energy Resources Inc.
1996	Chancellor Energy acquired by HCO Energy
1997	HCO Energy acquired by Pinnacle
1998	Pinnacle acquired by Renaissance Energy Ltd.
2002	Renaissance Energy Ltd. and Husky Oil Ltd. merged to form Husky Energy

3 THE NICHOLSON URANIUM MINE

3.1 Establishment of the Mine.

Probably the first comprehensive geological description of the Nicholson Bay area was given by Alcock in 1936 [14], with descriptions more specific to the Nicholson zone No. 4 deposit being given by others [8,17,76, 81-83,85,140-145].

The consolidated rocks in the Lake Athabasca region are all of Precambrian age, while those at the Nicholson site are mostly of Archean, early Precambrian age (Tazin Group), with traces of Proterozoic, late Precambrian age (Beaverlodge Series and Post-Beaverlodge). Those in the Tazin Group (shown in light blue in Figure 2.3) are generally limestone, dolomite, and secondary silicate rocks [14]. Those in the Post-Beaverlodge (shown in orange in Figure 2.3) are generally gabbro, peridotite, and amphibolite, while those in the Beaverlodge Series (shown in yellow in Figure 2.3) are generally quartzite, conglomerate, and iron formations [14]. The age of the pitchblende in the Beaverlodge area has been estimated at approximately 750 million years [146].

The Nicholson Bay area geology is described as a dipping sequence of bedded quartzite, dolomitic quartzite, and dolomite (see also Figure 2.6, above). Within this sequence was found a vein cutting the quartzite, containing pitchblende, carbonate, sulphides, and some gold. The zone No. 4 vein is described as striking N30°W, dipping at about 55° [83]. According to Griffith [8] and Beck [17] the Nicholson deposits are also known for their complex mineralization and large variety of minerals.

As noted in Section 2.3, encouraging trenching and drilling results were obtained in 1949. It is probable that ore-bearing rock was identified using a down-hole Geiger Counter and drill cuttings were collected and sampled for chemical analysis and ore grading[36]. As the zone Nos. 1 and 2 developments were not encouraging, these areas were shut down in favour of continuing work at the more promising zone No. 4 mine.

The encouraging results in zone No. 4, led to a mine shaft being sunk to 71 m, with principal lateral developments at the 30 m (100 ft.) and 60 m (200 ft.) levels [81] (Figure 3.1). It appears that a third lateral development was initiated at 100 m (325 ft.), but not further pursued [83]. Additional levels at about 150 and 180 m appear also to have been initiated but not developed, and the final depth of the shaft was about 213 m (700 feet) [147].

The 1949 lateral work showed that there were "rich shoots" in the zone No. 4 area, with "good width and grade" [83]. Discovery of a second "shoot" in an unexpected location suggested that other "branching or *en échelon* bodies" might exist [82], which provided enough encouragement to continue the development. By the end of 1950, about 466 m of lateral work had been completed at the first (30 m) level, and about 408 m at the second (60 m) level. At the upper level, several pockets of ore were found having grades of 0.38 and 3.75 percent (as U_3O_8), while at the second level the vein was reported to be "massive" [17,83].

Figures 3.1 and 3.2 show some of the zone No. 4 mine shaft development underway. Above the mine shaft a wood timber headframe[37] was built (Figure 3.3).

---

[36] An illustration of the processes of diamond drilling ore grading in this area and time period is given in a TMC documentary [61].

[37] Decades later, the headframe eventually became dangerously unsafe and it was intentionally burned down in 1991 [81]. See Section 6.2, below.

Figure 3.1. Early development of one of the Nicholson mine shafts in 1949. Courtesy of Saskatchewan Archives Board (Photos R-A10610-010, upper, and R-A9317-005, lower).

Figure 3.2. Miners loading a mine railway car, using a derrick crane, at one of the Nicholson Mine shafts in 1949. (Courtesy of Glenbow Museum Archives. Photo NA-4510-25).

Figure 3.3. Head-frame under construction at one of the Nicholson Mine shafts (probably at zone No. 4) in 1949. (Courtesy of Glenbow Museum Archives. Photo NA-4510-30).

3.2 Mining.

The main (zone No. 4) mine shaft was comprised of three compartments, two cage/skip compartments, and a personnel access, utility, and maintenance compartment [115]. The larger two compartments (see Figure 3.4) contained E. Long Ltd. cages suspended from a Canadian Ingersoll-Rand (C.I.R.) 107 cm by 76 cm (42 inch by 30 inch) double drum mine shaft hoist with a 75 h.p. motor, rope pull of 8,000 pounds, and 22 mm (7/8 inch) steel wire ropes.

The cages were identical, measuring about 249 cm high by 114 cm wide, by 145 cm deep (about 98 inch by 45 inch by 57 inch, see Figure 3.5). They had hinged, double doors on each face, and 18-inch gauge rails welded to their floors so that tram cars could be rolled in and out. The two cage compartments operated like elevators and could be loaded or unloaded from their front or rear faces. Either or both could transport miners and equipment up and down, or to lift tram cars of mined ore and waste rock ("muck") up to the surface. The third, "access" compartment was primarily for utility access (air, mine water, electrical, and ventilation) and maintenance, but would also have provided an emergency exit.

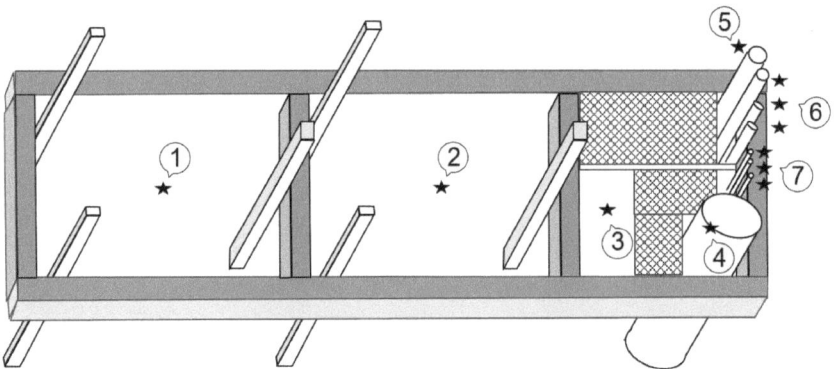

Figure 3.4. Illustration of the probable arrangement of underground mine shaft compartments, showing the cage/skip compartments (1 and 2), emergency exit (3), ventilation (4), high-pressure air (5), water supply and drains (6), and electrical conduits (7). Drawn based on specific information from reference [115] and period information from reference [68,95].

Figure 3.5. Photographs of the cage compartments. The upper-left image shows the pulley for the hoist rope. The right-hand image shows one of the hinged, double doors, and the rails to guide the tram cars can be seen at the bottom. (Saskatchewan Research Council, 2017).

It appears there were at least two ventilation "raises" in the main workings. Ventilation would have been provided by having fresh air drawn down through raises and workings, and then returning up the shaft. A small electric heater and fan were used intermittently on the top level during the coldest weather to prevent freezing in the shaft [115].

Two underground mining methods were used at Nicholson, *"open stoping"* with track haulage, and *"shrinkage stoping"* [115]. Stoping is a process wherein ore is removed from spaces or passageways called drifts, which then leaves open spaces called stopes. In stoping, miners would drill a pattern of holes, fill the holes with explosive, and then blast to break-up the ore[38]. The blasted, broken-up ore could then be mined and transported. These methods were possible because the surrounding granite rock was stable enough not to collapse after the ore had been mined out, although the stopes would have been stabilized with timbers bolted to the rock.

In overhand open stoping, the deposit is mined upwards from the

---

[38] An illustration of the process of underground uranium mining in this area and time period is given using film footage from the nearby Rix Athabasca mine in a TMC documentary [61].

bottom, removing the broken-up rock as mining progressed. This was done where the orebodies were fairly horizontal, and it was completed during the winter months, at which time the ore was lifted out of the mine and shipped to the Eldorado mill. Shrinkage stoping was conducted in the steep-dipping veins and involved blasting and mining upwards, in horizontal layers, from a lower horizon to an elevated horizon. The broken-up rock was then allowed to fall and remain in-place at the bottom of the stope. In this case, the broken rock provided both a working platform and a means to help stabilize the stope[39]. Shrinkage stope mining was conducted during the summer and fall months, leaving the ore to be reclaimed during the winter, along with the open-stope-mined ore [115].

Diamond drilling[40] was completed by two-miner crews "normally drilling and blasting about twelve 8-foot holes per shift" [115]. Access-ways and working platforms were constructed from planks and ladders, supported by steel bolts drilled into the walls. Ore shoots were located by diamond drilling upwards at 25-foot intervals and then testing the recovered cores (or sludge) radiometrically. Two muck samples were taken from each car hoisted to the surface, and tested radiometrically. Common practice in the region was to back-fill mined-out drifts with waste rock to further enhance stability, although it is not clear whether this was done at Nicholson.

According to Mamen [115] a slusher, comprising a double drum hoist and a scraper, was used to drag ore from drawpoints and pillar stopes. In this method, one cable was attached to the front of the scraper bucket so that it could be pulled along a drift, while a second cable would be anchored at the drawpoint or pillar used to return the scraper bucket for another load. The Nicholson mine had both electric (C.I.R. A5 NNOH) and compressed-air (Gardner-Denver HKE) driven double-drum (slusher) hoists, driving 0.66 and 0.86 m (26" and 34") Pacific Slushmaster scrapers [115].

In terms of access ways, drifts and cross-cuts were about 1.8 m by 2.1 m (6 ft by 7 ft) and the raises were 1.5 m by 1.5 m (5 ft by 5 ft) [115].

On the main underground levels, ore and waste rock were transported in tram cars, propelled by a compressed air tram called an "Air Trammer," (Figures 3.6 and 3.7) which moved on a narrow-gauge railway having 18-inch gauge rails. The tram cars had approximately 0.5 and 0.6 m^3 (18 and 20 cu ft) capacities, and were of a swiveling, end-dump design (Figure 3.8). The tram cars were loaded using a rail-mounted compressed-air rocker-shovel (Eimco 12B Loader), whose front-mounted shovel could load from the

---

[39] However, some of the blasted rock has to be removed from the drift as mining progresses, because broken-up rock takes up more volume than it did when intact.

[40] The drills used were Copco and C.I.R. airleg machines; and Copco machines with extension legs for the stoppers.

front or sides of the track and then throw its contents into an attached tram car. For more details, see reference [115]. The ore and waste rock were transported by rail along the haulage levels to the main shaft for hoisting to the surface (Figure 3.9 and Appendix 7).

Figure 3.6. Air trammer (left) and tram car (centre) similar to those used at the Nicholson Mine. The ones shown are at the nearby Nesbitt-Labine Uranium Mine, circa. 1952. Courtesy of University of British Columbia, Rare Books and Special Collections (Northern Miner fonds).

Fresh water was drawn from Lake Athabasca and piped down the mine shaft. Controlling the ingress of water from seepages would have been the greatest water problem in the mine. Such seepage was a common problem experienced by all the uranium mines in the region. As mining progressed, drilling and blasting would have constantly exposed ever more fissures and cracks from which water could seep into the mine, threatening to flood the drifts and shaft. The Nicholson mines were no exception. Lang [78] reported that *"During the early stages of the lateral work, great difficulty was experienced because of strong flows of water from vugs and caverns, but this water was finally drained."* This was accomplished by constant pumping out of the water using a 25 h.p. Cameron 2MRV pump and a 30 h.p. Berkely 2½ Zp Series A pump [115]. These had a combined pumping capacity of about 3,000 litres per minute (550 gal per minute). The general mining practice in the mid-1950s was to have automatic level-sensing controls for the pumps, set to maintain the water levels in the drifts at no more than 0.6 to 1.2 m (2-4 ft) from the bottoms [28].

Cold was also an issue for all of the northern mines, and heat emanating from the mine machinery, bolstered by underground heaters fueled by Bunker heating oil, combined with circulation aided by fans, were generally utilized to maintain practical working conditions [28].

Figure 3.7. The chassis from the Nicholson air trammer in 2017 (Saskatchewan Research Council).

Figure 3.8. Tram cars sitting outside the Nicholson mine in 2017. The lower photo shows the swivel mechanism. (Saskatchewan Research Council)

In Section 1.4 above, it was noted that the health effects of low doses of radiation were not yet well known in the 1950s, so uranium mining was still considered "safe" (as far as radiation hazards were concerned). In the television documentary film, "*The Road to Uranium*" [63], a reporter speaks to one of the underground uranium miners at the nearby Eldorado mine in 1957, and asks "*Is uranium dangerous to work with? Do you have to be careful?*" to which the miner replies "*No ... [I] never experienced anything anyway.*"

Underground uranium mining was hazardous, however. A constant flow of air had to be provided, not just for breathing, but to dilute the exhaust fumes of diesel-powered mining equipment, blasting gases, along with the dust particles (both radioactive and non-radioactive) created by the mining itself. Although the hazard from breathing radioactive particles was not yet well known, the risk of silicosis from breathing silica particles was understood [28]. Air pressure, fans, and large ventilation pipes were used to ventilate the mine workings, but not always very effectively. An examination of one of the nearby Eldorado Beaverlodge mines (probably Ace or Fay) in 1954 identified "*active dead ends at the mine with no ventilation*" [28].

By the end of 1950, considerable lateral work had been completed at two primary levels, one at 30 m (100 ft) and another at 60 m (200 ft). As mining progressed at these two primary levels, the shaft was also deepened to about 107 m (350 ft.), while some lateral work was undertaken at the 99 m (325 ft.) level [81]. Figure 3.9 shows an isometric view of the workings at the two primary levels. (See also Appendices 7 – 10.)

As noted previously, shrinkage stoping was conducted during the summers, with the ore being stockpiled for shipment in the winters. In the winters, both shrinkage stoping and open stoping were conducted. The winter deliveries of ore to Eldorado were typically between the beginning of January and as late into spring as the ice thickness could still support the weight of the ore-carrying trucks over the 29 km (18 mile) route over Lake Athabasca and Beaverlodge Lake. These winter shipments were at a rate of about 227 tonnes per day (250 tons per day) [115], and occurred from 1954 through 1956 [121,132,133,148-151].

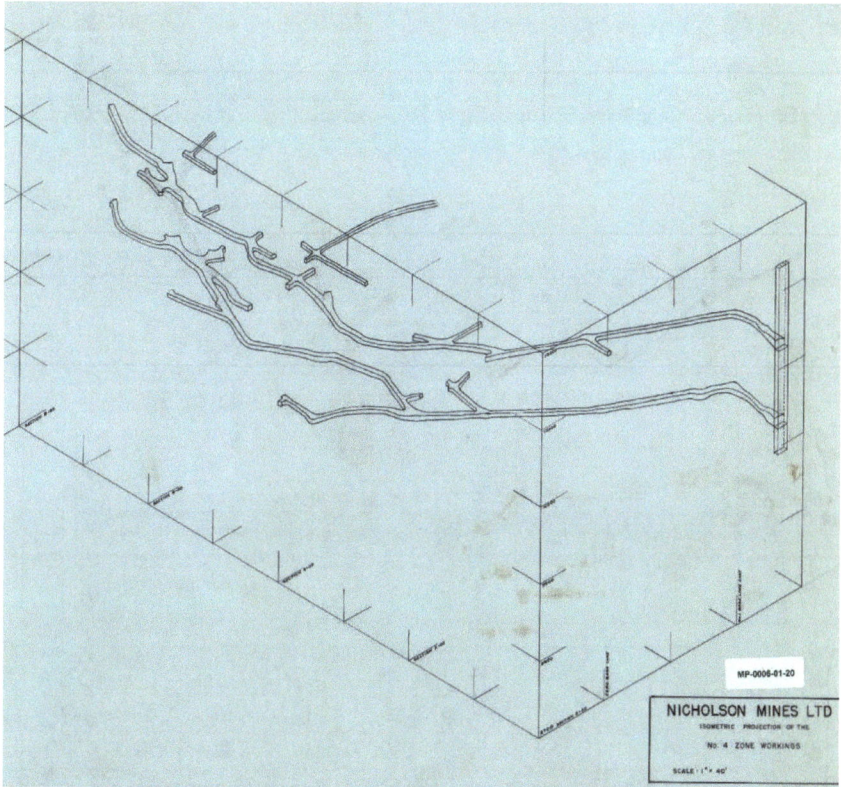

Figure 3.9. Section from an isometric drawing of zone No. 4 workings (Nicholson Mines Ltd., circa 1958). The complete drawing is shown in Appendix 7.

3.3 Waste Rock.

Waste rock was produced from each of the main shaft #1 in zone No. 4, and from the secondary, exploration shafts #2 in zone No. 2 and #3 in zone No. 1. These produced four substantial waste rock piles (Figure 3.10).

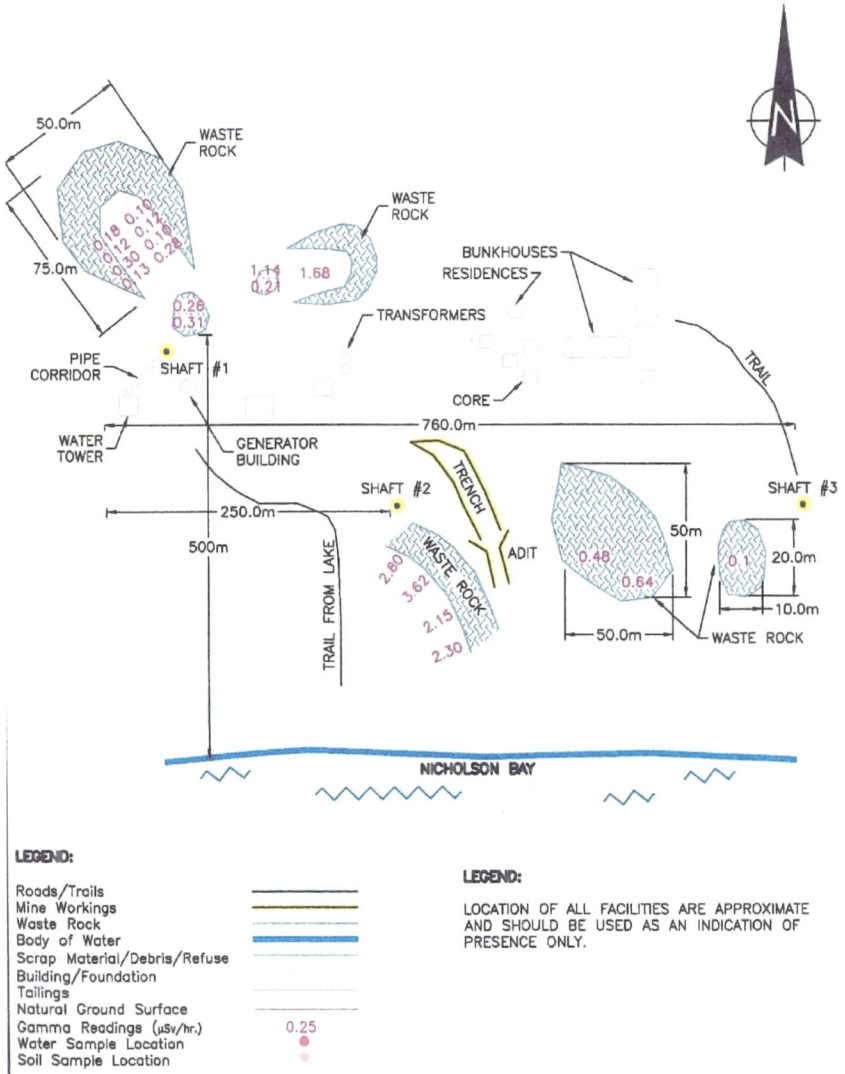

Figure 3.10. Consolidated Nicholson Uranium Mine Site Plan (Illustrative sketch map only; not to scale). Reference [81].

The waste rock was placed (mostly) in two large piles near the main mine shaft #1 (Figure 3.10), one west of the shaft about 75 m by 50 m, by 20 m high, and estimated to contain 50 to 75 thousand m³ of dolomitic and quartzitic rock[41](see Figure 3.11), and a smaller one nearby estimated to contain another thousand m³ [81]. A third waste rock pile was created near shaft #2, and is about 50 m by 50 m, by 5 m deep, with an estimated volume of 10 to 15 thousand m³ [81]. Finally, a fourth waste rock pile was created near shaft #3, and is about 10 m by 20 m, by 1 m deep, with an estimated volume of 200 m³ [81].

In total, about 60 to 90 thousand m³ of waste rock was left near the three shafts on the site [81].

Figure 3.11. The main waste rock pile (Saskatchewan Environment, 2001 [81]).

Most of the ore sent from the Nicholson mine to Eldorado for milling was about eleven thousand tonnes grading at 0.3% (as U_3O_8) in the early 1950s, and about two thousand tonnes grading at 0.5% (as U_3O_8) in the late 1950s [16,17,81,84]. Reports vary regarding the total amount of uranium ore mined from the Nicholson mine during its operating years, but most

---

[41] This rock also contains some copper, which may be relevant to the remediation process.

conclude that it ultimately yielded approximately 50 tonnes of yellow-cake (as U_3O_8) – see Table 3.1.

In summary, then, the Nicholson Mine is thought to have produced a total of about thirteen thousand tonnes of uranium ore, that would have yielded about fifty tonnes of yellow-cake uranium (as U_3O_8) after milling by Eldorado.

Table 3.1. Estimates of the Uranium Produced from the Nicholson Mine During its Operating Years.

Total Uranium Produced 1949-1959 (tonnes U_3O_8)	Reference
54.4	Hassard, for Imperial Oil Ltd., 1975 [137]
48.3	Saskatchewan Geological Survey, 2003 [16]
48.3	Jensen, for Red Rock Energy Inc., 2005 [85]

4 MINE INFRASTRUCTURE AND THE NICHOLSON CAMP SITE

The Nicholson site was (and still is) somewhat remote, existing in a sub-arctic region that is semiarid with short cool summers and cold winters. Being located near the shore of a large lake in Saskatchewan's north, the site was subject to almost constant winds. Temperatures ranged from cool in the summer, (averaging 17 °C in July, with a maximum recorded high in the region of 35 °C in 1984), and very cold temperatures in the winter (averaging -27 °C in January, with a maximum recorded low in the region of -49 °C in 1974) [66,152].

The uranium companies in the region were able to attract a workforce to remote locations such as this only by paying close to the highest wages in the mining and mineral industry in Canada [153].

In his biography, Nicholson is quoted as relating that in 1930 they built the first cabins on the site to enable full-time seasonal work on the property, and that they spent the summer of 1930 "blasting into the showings" [90].

The nearby village of Goldfields was short-lived. It was established in 1937, and by 1939 had a population of over 1,000 [43,154]. However, when the Box gold mine and mill ceased production in 1942, the village closed as well. At this point, some buildings were torn down, with the lumber being reclaimed and shipped to Fort McMurray [43]. It appears that some buildings remained in use, with people staying on in the village, particularly when the local wave of uranium exploration activity in the area began in 1944.

When significant work was re-established at the Nicholson mine site in 1948 and 1949, access continued to be restricted to travelling by air or over water (or ice). The closest settlements were now the village of Goldfields about 3 km (2 miles) to the west, and the town of Eldorado about 11 km (7 miles) to the northwest. Neither of these were very convenient to access nor very stable. As a result, more permanent buildings were constructed at the Nicholson site (see Figures 4.2, 4.9, and 4.13 – 4.16, below).

The town of Eldorado was established in 1951, when the company's Fay uranium mine was being constructed, and this town quickly grew to house about a thousand people.

In 1952, the provincial government decided to establish Uranium City, with the support of Eldorado Nuclear Ltd., several kilometres away from the town of Eldorado. This caused the demise of both the village of Goldfields and the town of Eldorado. The remaining salvageable buildings from Goldfields, including the school, hotel, along with numerous residences, were dragged across the ice on Lake Athabasca to Uranium City in 1952 and 1953 (Figure 4.1) [28,37,43, 154]. Some descriptions of Uranium City and life in the town in this time period are given in references [37,43-46,155], and illustrated in the 1953 TMC [60] and 1957 ITN [63] documentary films.

Figure 4.1. Relocating a house from Goldfields to Uranium City in 1953. Courtesy of University of British Columbia, Rare Books and Special Collections (Northern Miner fonds).

The communities of Goldfields, Eldorado, and Uranium City, in their respective eras of operation, would have provided sources of general supplies, but during the Consolidated Nicholson mine's peak operating years there was only Uranium City, some 16 km away. As a result, the company built its own dedicated infrastructure onsite, with some buildings near the headframe and others near the shore of Nicholson Bay (Figure 4.2).

As noted in Chapter 2, the preferred, and most economical, option for large-volume or high-mass transportation was to use a barge service. However, seaplane and ski-plane services were available throughout most of the year, and these would have been used for urgent transportation of everything from construction materials and general supplies, to emergency parts and groceries.

Figure 4.2. Aerial view of buildings near the shore of Nicholson Bay in 1981 (Saskatchewan Energy and Resources).

4.1 Mine Infrastructure.

The surface plant comprised a timber headframe and shaft-house, a hoist room, and a boiler house with an oil-fired boiler that was used to heat the shaft-house and headframe [115] (see Figures 3.3, 4.3, 4.4, and also Appendices 1 - 5).

Beyond the mine workings and headframe was a rail tramway from the mine. Other mine infrastructure included a power house, electrical sub-station, and auxiliary power house, a water tower, mine office, core shack, and powder magazine, separate blacksmith and carpenter/machine shops, a warehouse, and a garage.

Figure 4.3. Head-frame construction at the Nicholson Mine site in 1949. (Glenbow Museum Archives. Photo NA-4510-26).

Figure 4.4. The boiler house, beside the head frame in 1970 (Courtesy of Ben McIntyre).

Water and Fuel. Most, if not all, bulk petroleum products, including diesel oil and gasoline, would have been shipped to Nicholson during the open-water season each year. Diesel oil was stored in a 318 kL (70,000 gal) main tank and four 13.6 kL (3,000 gal) tanks [114]. A water tower was constructed on a hill near the headframe and boiler house (see Figures 4.5 and 4.6).

Figure 4.5. Early development of the Nicholson mine site in 1949. This appears to be construction of the base for the water tower on zone No. 4. Courtesy of Saskatchewan Archives Board (Photo R-A9317-007).

Figure 4.6. The water tower in 2001. (Saskatchewan Environment [81]).

Power. Electric power was purchased, when available, from Eldorado who in turn had leased the Wellington Lake hydroelectric plant[42]. This operation, which is about 25 km west of Uranium City, involved diverting water from Tazin Lake to White Lake. White Lake had been dammed, and its water used to feed a power plant at Wellington Lake [93]. However, this source of power was not reliable for the Nicholson operations as they could only buy "surplus" power, when it was available, from Eldorado [65]. For this purpose, the Nicholson site had incorporated an electrical substation (Figures 4.7 and 4.8).

Diesel power was used when the hydroelectric power was unavailable or insufficient for the Nicholson mine and camp operations. For this, the site had an auxiliary powerhouse with a 220 h.p., 125 kW Paxman diesel generator and an 80 h.p., 50 kW Ruston Hornsby diesel generator [115].

Figure 4.7. The electrical sub-station in 2001. (Saskatchewan Environment [81]).

----

[42] The dam, hydroelectric generating facilities, and transmission line were originally built between 1935 and 1938 by Cominco for the Box gold mine, then later expanded so it could supply power for Eldorado.

Figure 4.8. Transformers outside the electrical substation (Saskatchewan Research Council, 2015).

Compressed Air. Several mining machines, such as the air tram and one of the double-drum (slusher) hoists, operated on compressed air. For this, the site had a Leroi 500 D2C compressor powered by a Caterpillar D-13000 diesel generator, and a C.I.R. XVHE-2 compressor powered by a 200 h.p. electric motor, the latter compressor being used when hydroelectric power from the Wellington Lake station was available [114].

Mine Operations. Operating infrastructure included a mine office (Figure 4.9), a core shack, separate blacksmith and carpenter/machine shops (Figures 4.10 and 4.11), and a garage. Due to the remote location of the site, and the short seasons available for shipping by waterways a warehouse was used to maintain stocks of spare parts and consumable goods, including oil, and sulphur, with a separate powder magazine for explosives.

Rolling stock for the site, as of 1956, included a D4 Caterpillar bulldozer, TD 9 International bulldozer, Willys Jeep, Mercury 3-ton truck, and a Chevrolet ½ ton pickup [115]. Ore haulage from the mine to Eldorado was handled by a contractor using six trucks, loading 7-9 tonnes per load, and making six trips per working day of 14 to 16 hours [115].

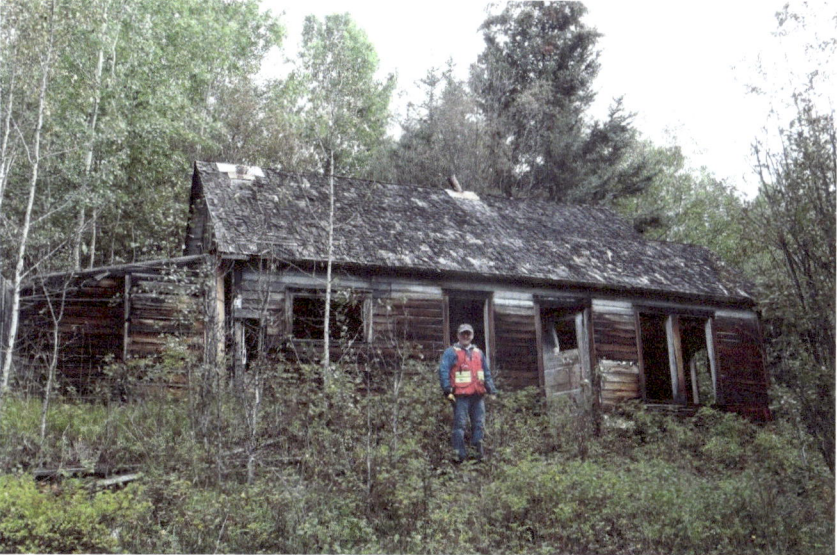

Figure 4.9. The mine office building, near the lakeshore, in 2017. (Saskatchewan Research Council)

Figure 4.10. The blacksmith shop exterior (above) and interior (below) in 2015 (Saskatchewan Research Council).

Figure 4.11. The machine shop exterior (above) and interior (below) in 2015 (Saskatchewan Research Council).

4.2 Camp Infrastructure.

John D. Nicholson's biography [90] refers to their first camp at the Nicholson mine site being essentially a collection of tents. Log-cabin structures were built later, probably in the early- to mid-1930s (Figure 4.12).

When Nicholson resumed prospecting and development work in 1948 and 1949, additional accommodations would have been required for a growing workforce, which comprised about 70 people in 1949, and about 35 additional people in 1950 [156]. In 1948 and/or 1949 a 40-person bunkhouse and a cookhouse were constructed (Figure 4.13 and 4.14). The photograph in Figure 4.15 shows that by late 1949 there were at least 15 structures of various sizes in place (see also Appendices 1 - 5). Figure 4.16 provides another example.

Figure 4.12. Early development of the Nicholson camp. Although this photograph was taken in 1949, the log cabins probably date from the early- to mid-1930s. Note the suspended water barrels for taking showers. Courtesy of Saskatchewan Archives Board (Photo R-A9317-007).

Figure 4.13. The bunkhouse in 1949. Courtesy of University of British Columbia, Rare Books and Special Collections (Northern Miner fonds).

Figure 4.14. The cookhouse (middle structure) in 1949. Courtesy of University of British Columbia, Rare Books and Special Collections (Northern Miner fonds).

Figure 4.15. Aerial photographs showing the state of the camp infrastructure by late 1949. Courtesy of University of British Columbia, Rare Books and Special Collections (Northern Miner fonds).

Figure 4.16. One of the residences near the lake shore. (Saskatchewan Research Council, 2015.)

The camp infrastructure included:

- The bunkhouse and at least four other residence buildings,
- Cookhouse,
- An auxiliary power house,
- A food warehouse and refrigeration unit, plus a dry storage building,
- A washhouse, and a steam-bath house.

The water supply was from the lake, and involved pumping the water into 2,000 gallon wood-stave tanks located in each building [115]. The buildings were heated independently, using oil-fired space heaters [115].

In his autobiography [135], Ted Kennedy provides a description of the Nicholson camp as experienced by he, his wife Celia, and their three children in January of 1959:`

"We all pitched in and had the camp, which consisted of two two-story bunkhouses and six houses, rehabilitated, setting up an office/cookhouse, and three family dwellings ... We moved ... to a small house at Nicholson. A kitchen, bathroom, living room and bedroom downstairs, with two small bedrooms upstairs, although one was filled with four 45 gallon water drums which we pumped full from the lake once or twice a week. The plumbing was hooked on to these so as long as we had water in the drums upstairs, there was water in the taps down ... It's cold in the winter in that

country, and that winter was no exception. It got to be 60 below on a few nights and 50 below was common."

The recreational activities in this area, and era, generally included boating, hunting (such as for caribou, deer, and moose), and fishing (for trout, jackfish, and grayling) [66,89,157]. Kennedy also provides an anecdote about Springtime at Nicholson in 1959 [135]:

"As the days grew longer and weather warmed up, … Spring finally arrived and everybody was in good spirits. The leaves came out, the grass turned green and the animals awoke from their long sleep. Ducks and geese returned from the south. A great feeling of joy and anticipation invaded the camp … One day the miners were mining, the cook was cooking, the mechanic was mechanicing [sic], and the truckers were trucking. The weather was beautiful. The kids were playing in the sandpile behind the house. Celia was in the house doing the dishes at the sink by the kitchen window. Suddenly, quietly, along the path strolled a big old black bear – right out where [their children] were playing in the sand. Celia took one look, hollered some unprintable obscenity, and was out the door in a flash, yelling and screaming at that bear and throwing pots and pans, tin cans, rocks and what have you to get him away from the kids. The bear, who hadn't even noticed the boys, took off like a scared rabbit and was never seen again. He probably spent the rest of his life hiding somewhere in northern Manitoba."

5 MILLING AND REFINING

Although Consolidated Nicholson Mines decided that their uranium reserves did not warrant building a dedicated mill in the Beaverlodge area, Eldorado did (Figures 5.1, 5.2).

In 1951, Eldorado made the decision to build a 454 tonne per day (500 ton/day) capacity uranium mill near the Ace-Fay mines, in the Beaverlodge area (with construction commencing in 1952 and production starting in the spring of 1953) [10]. It was expanded to 680 tonne per day (750 ton/day) in 1955, and then again to about 1,800 tonne per day (2,000 ton/day) in 1957 [158]. It was to the Eldorado mill that all of the Nicholson ore was sent for milling, between 1955 and 1959.

Figure 5.1. The Eldorado Beaverlodge mill building in 1955. Courtesy of Saskatchewan Archives Board (Photo R-A5428).

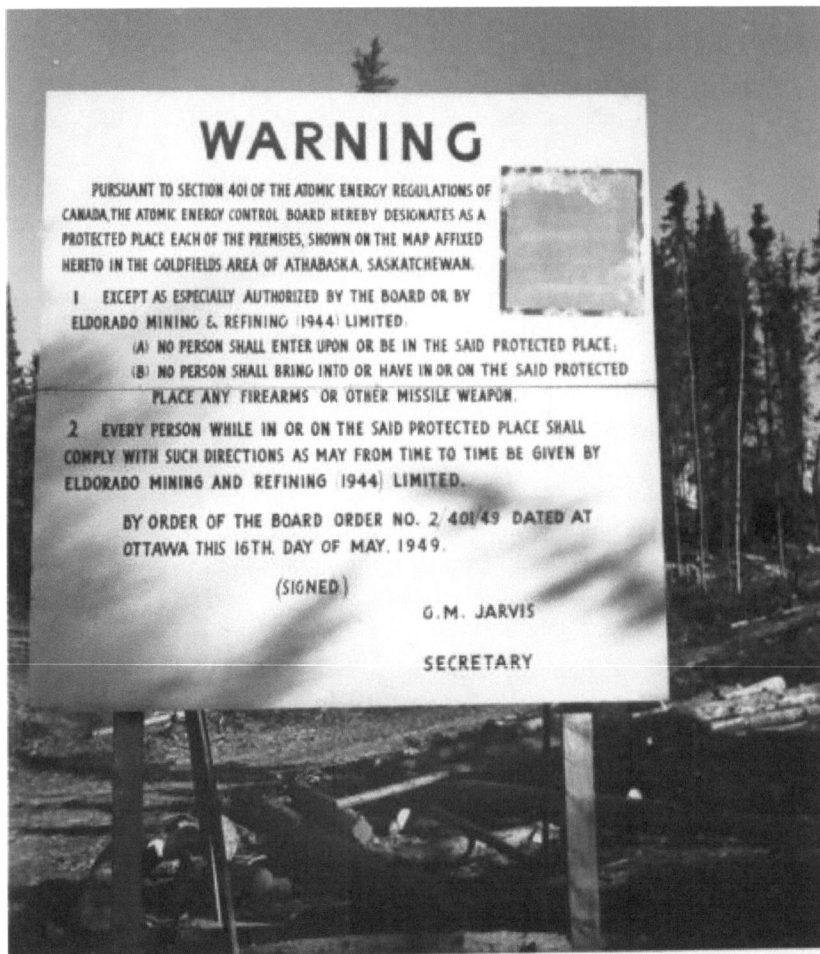

Figure 5.2. Sign designating the Eldorado mill building as a protected area, 1949. Courtesy of Saskatchewan Archives Board (Photo R-A9252).

5.1 Uranium Milling Background.

The Atomic Age and Cold War Era uranium mills were somewhat unique in Canada in that they were in very remote locations (making the shipping of process equipment and chemicals to the mills unusually expensive), with harsh winter climates (creating special problems of heating, moisture, and sometimes permafrost), they were required to efficiently process low grade ores (for which an alternative to the classical "gravity method" of separation had to be developed and then adapted to the particular ore from a given mine), and they required "EMF control" (to maintain a particular oxidation state on the part of the uranium being leached). The Canadian uranium ores contained pitchblende and uranophane with a large percentage of uranium (IV) [1], so the mills had to control the oxidizing nature of the leach solutions.

Most of these issues were dealt with in the development of the first two large uranium mills in Canada, namely those at Port Radium and Beaverlodge [1]. The Canadian Department of Mines and Technical Surveys (part of what is now Natural Resources Canada) developed an alkaline leach process that was introduced at Port Radium in 1952, and then adapted for and introduced at Beaverlodge in 1953 [1]. These were the first two commercial-scale, continuous plants for leaching uranium ores in the Western Hemisphere (and the second and third in the world after one in Portugal) [1].

The reason for using an alkaline leach process, rather than the more conventional (at the time) acid leach process, was the high carbonate content of ores such as those that were going to be produced at Eldorado's Ace Mine. High carbonate-content ore would cause high acid consumption in an acid leach process, so it seemed that an alkaline process might be better, but the company was not confident about this until pilot testing was completed at the Sherritt Gordon facility in Ottawa [10]. Ultimately, the alkaline leach process was adopted for the Eldorado mill, which commenced operations in 1953 (conversely, an acid leach process was developed for the nearby Gunnar mill [66,68,153]).

5.2 Uranium Milling at Eldorado - Beaverlodge.

The Eldorado milling process consisted of a number of operations, which have been described in some detail in the public literature [1,8,15,65,66,107,158-160]. A simplified process diagram for uranium ore milling at Eldorado's Beaverlodge, Saskatchewan mill is given in Figure 5.3. The extraction efficiency of the process averaged about 90% (as U_3O_8) of the uranium in the ore introduced into the plant [158].

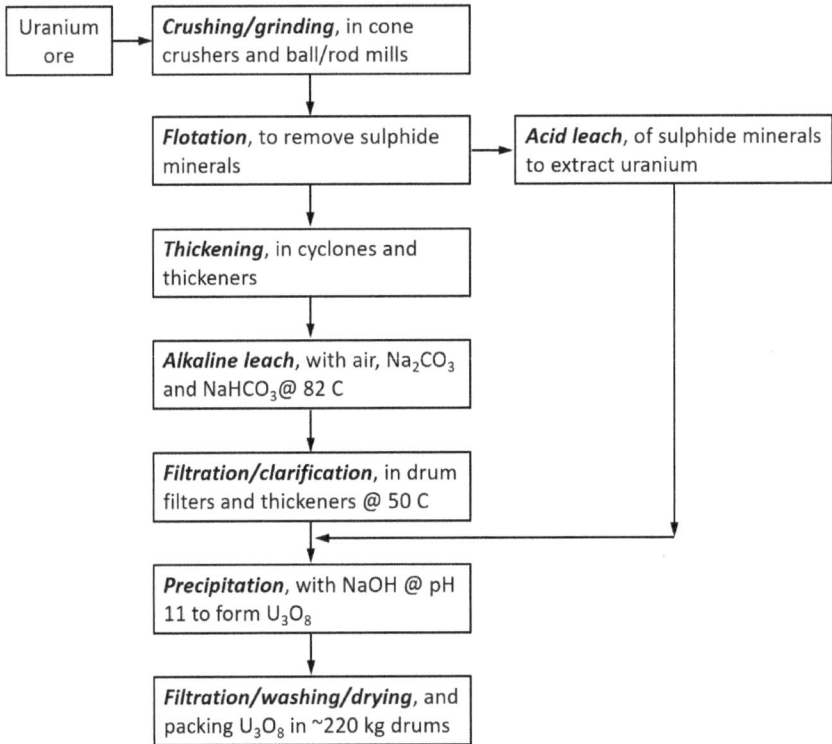

Figure 5.3. Simplified process diagram for uranium ore milling at Eldorado's Beaverlodge, Saskatchewan mill.

Crushing and Grinding Circuit. The raw ore was first sorted using a scintillation counter, such that only ore lumps with greater than a specified level of radioactivity would be transferred to the crushing and grinding circuit. The ore was then crushed through a succession of cone crushers and ball mills that progressively reduced the largest sizes down to mostly less than 0.074 mm (0.0029", 70% minus 200 mesh) – Figure 5.4. The fine solids were then sent to a flotation circuit for the removal of sulphides.

Figure 5.4. The crushing and grinding circuit in the Eldorado Beaverlodge mill *circa* late 1950s. Courtesy of University of British Columbia, Rare Books and Special Collections (Northern Miner fonds).

Flotation Circuit. The solids were slurried to about 26 mass%, and excess sulphides (above 0.2%) were removed by flotation, at pH 10, in two stages using isopropyl xanthate as a flotation collector and Dowfroth as a frother. Following "rougher" flotation in a bank of Denver flotation cells, the rougher concentrate was subjected to a second stage of "cleaner" flotation in another bank of Denver flotation cells. The floated sulphide material was sent to a batch-mode acid leaching tank (40 ton/day capacity) in which the uranium was leached out. The flotation concentrates were thickened to 50 – 55 mass% using cyclones and thickeners, and then pumped to the leaching circuit.

Leaching Circuit. The uranium was leached out of the ore using four production lines, containing six "*pachuca*[43]" leaching tanks per bank (Figure 5.5). *Pachucas* are essentially large process vessels in which ore slurry and air are mixed by countercurrent flow. Each *pachuca* vessel was about 8 m (26 ft) in diameter, about 15 m (50 ft) high, and having a 60° cone bottom.

Key to the alkaline leach process is the dissolution of the uranium and the stabilization of it in solution as uranyl tricarbonate ions. This requires all of the uranium to be in the oxidized hexavalent ("uranyl") state, so the fraction of the ore having uranium present in its tetravalent ("uranous") state has to be oxidized first. At the time the Nicholson ore was processed, Eldorado was using a central air lift, injecting compressed air through eight separate diffusers in each *pachuca* (probably having nozzle diameters of about 2.5 cm [161]), to provide both agitation and oxidizing conditions. (In later years, it appears that sodium chlorate additions were introduced to increase the oxidizing conditions, and in yet later years oxygen was also introduced directly into the vessels to further increase and/or maintain the oxidizing conditions.) This process was also aided by maintaining a relatively high process temperature of 82 °C (180 °F), for which the heat was provided by steam coils.

With the uranium (IV) oxidized to uranium (VI), all of the uranium would be soluble in water. In the presence of sodium carbonate (soda ash, Na_2CO_3), the overall chemical reactions for uranium (IV) and uranium (VI) oxides in the pitchblende ore were as follows.

For U(IV):

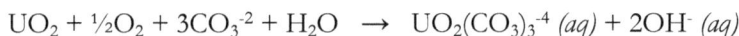
$$UO_2 + \tfrac{1}{2}O_2 + 3CO_3^{-2} + H_2O \;\rightarrow\; UO_2(CO_3)_3^{-4} \; (aq) + 2OH^- \; (aq)$$

For U(VI):

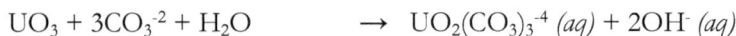
$$UO_3 + 3CO_3^{-2} + H_2O \qquad \rightarrow \quad UO_2(CO_3)_3^{-4} \; (aq) + 2OH^- \; (aq)$$

To maximize uranium dissolution, the excess hydroxyl ions were probably neutralized with bicarbonate:

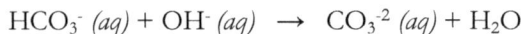
$$HCO_3^- \; (aq) + OH^- \; (aq) \;\rightarrow\; CO_3^{-2} \; (aq) + H_2O$$

The residence time for ore in the leaching circuit was about 96 hours, and the extraction efficiency was in the range of 90-92%.

---

[43] These are named for Pachuca, a city in Mexico that is famous for its long-standing silver and gold production. The origin of the name may be the Spanish term *pachoacan* (place of silver and gold).

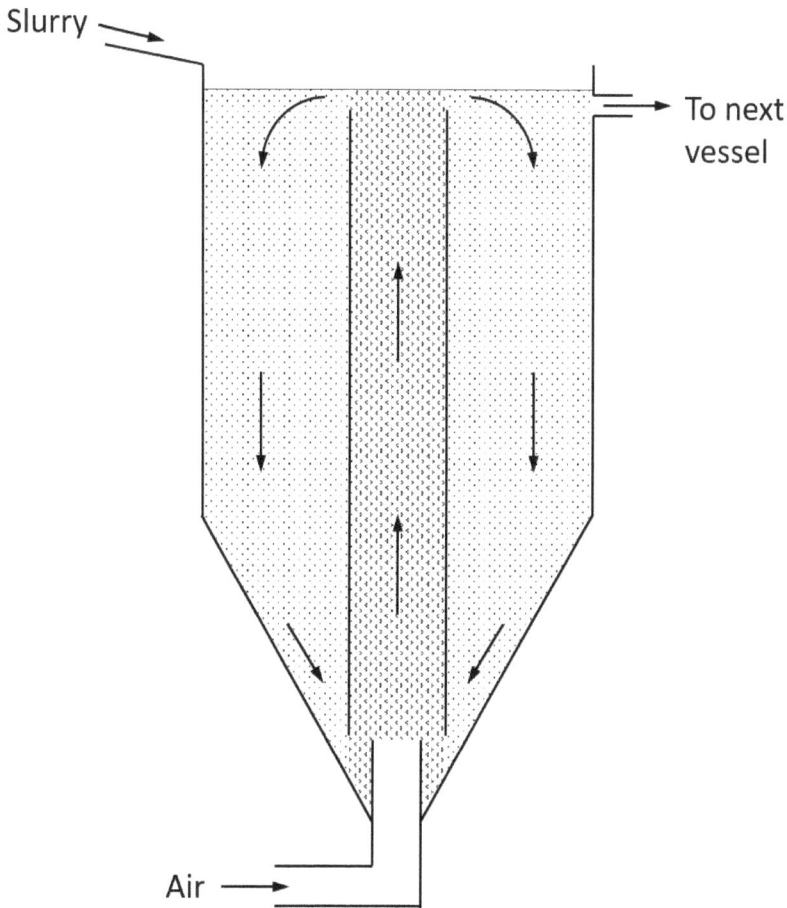

Figure 5.5. Simplified illustration of one of a series of *pachuca* vessels, with an air-injection draft tube to disperse and lift the slurry up the centre of the vessel.

Filtration and Clarification Circuit. Filtration was used to separate undissolved solids from the dissolved uranium. In each production line the ore slurry was pumped from the sixth *pachuca* leaching tank via a concentric pipe heat exchanger, which reduced the temperature from 82°C to about 50 °C while transferring heat to the wash water. The filtration was accomplished using string-discharge drum filters. The filter cakes on the drums were washed with warm water, re-slurried, and re-washed[44], and then

---

[44] The filtration residues were used for backfilling mined-out stopes in the nearby Eldorado mine.

treated with lime to remove excess bicarbonate ions and flowed to a thickener tank for clarification by sedimentation. The resulting clarified solution was referred-to as "pregnant solution."

Precipitation Circuit. Dissolved uranium ions were precipitated from the "pregnant solution" in a series of six agitators, using sodium hydroxide[45] (caustic soda) to form a solid uranium oxide concentrate that is commonly known as "yellowcake" – $Na_2U_2O_7$. The precipitation reaction was:

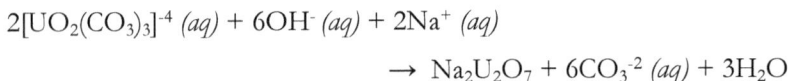

$$2[UO_2(CO_3)_3]^{-4} \; (aq) + 6OH^- \; (aq) + 2Na^+ \; (aq)$$
$$\rightarrow Na_2U_2O_7 + 6CO_3^{-2} \; (aq) + 3H_2O$$

This process was probably maintained at about pH 11. At high pH, and with oxidizing conditions no longer being maintained, some uranium (VI) would have become reduced to uranium (IV) during this process. Yellowcake has the approximate chemical formula U_3O_8, is actually a mixture of uranium (IV) and uranium (VI) oxides, and is usually approximately $2/5$ UO_2 and approximately $3/5$ UO_3.

Product Filtration, Drying, and Packaging. The precipitated yellowcake was filtered in plate-and-frame type filter presses that enabled batches of the yellowcake precipitate to be washed and filtered. The filter cake was then dried for packaging in 113 litre (25 imperial gallon) steel drums (see Figure 5.6). These drums weighed about 204-227 kg (450-500 lb) each. The final uranium precipitate was about 79% U_3O_8.

Eldorado Mining and Refining had constructed an airstrip at Beaverlodge in 1951 [66]. This enabled them to fly the drums of yellowcake directly from the mine airstrip to Edmonton by an Eldorado Aviation aircraft (Figure 5.7), and from Edmonton the drums were shipped by rail to Port Hope, Ontario [1,10,107].

---

[45] Possibly with the addition of magnesium oxide (MgO) as well.

Figure 5.6. Employees at the nearby Gunnar mine and mill pose with their first barrel of yellowcake in 1957. Courtesy of University of British Columbia, Rare Books and Special Collections (Northern Miner fonds).

Figure 5.7. Loading a barrel of yellowcake, from the nearby Gunnar mine and mill, into an Eldorado Aviation DC-4 aircraft (CF-JRW), for shipment to Port Hope via Edmonton. Gunnar Mines Ltd., 1958 [162].

5.3 Uranium Refining at Eldorado – Port Hope.

The yellowcake produced from all ores sent to the Eldorado mill at Beaverlodge was refined at Eldorado's refinery in Port Hope, Ontario (Figure 5.8). Although this refinery had been in operation since 1933, its initial focus was on the production of radium and uranium was simply a by-product. By 1942, the plant's focus had begun to shift to uranium and radium production, and by 1953, the radium production was discontinued, with the plant focusing on just uranium production [8].

Figure 5.8. The Eldorado refinery, Port Hope, in 1938. Library and Archives Canada (Photo 3375914).

In this era (1940s through the mid-1950s), the refining process apparently involved dissolving the uranium in nitric acid and then extracting the uranyl nitrate (twice) into ether, then re-extracting it into water (with very large volumes of water), and, finally, heating ("calcining") the purified uranyl nitrate to oxidize the uranium and produce uranium trioxide [65,163]. Accounts vary, with regard to the appearance of the final product, but it seems likely that the original refining process produced "black oxide" (about 95 percent U_3O_8), but that by the early 1950s the product was likely "orange powder," a fairly pure form of uranium trioxide (UO_3).

In the early 1950s, the details of the uranium refining processes used in Canada and the U.S. were held secret [151], although it was known that yellowcake was being processed to yield "orange powder" (uranium trioxide, $\gamma\text{-}UO_3$). In some refineries the uranium trioxide was being reduced to "brown powder" (uranium dioxide, UO_2), and then converted to "green salt" (uranium tetrafluoride, UF_4) [151,164] (see Figure 5.9). In the U.S., the uranium tetrafluoride was then converted to uranium metal and/or uranium hexafluoride (UF_6).

Figure 5.9. Some forms of uranium: yellowcake ($Na_2U_2O_7$, left), green salt (uranium tetrafluoride, UF_4, centre), brown powder (uranium dioxide, UO_2, right), and uranium metal (lower). (Canada Department of Energy, Mines, and Resources [8]).

Since the late 1950s, however, more information has become publicly available such that several descriptions of the Eldorado refining process are available [1,8,151,163,165]. Only a summary of the process (*circa.* 1955 through 1959) is provided here.

By the time the first yellowcake from the Consolidated Nicholson mine would have arrived via Eldorado's Beaverlodge mill to the Eldorado refinery at Port Hope, in 1956, the refinery had converted to a simpler solvent extraction process than the one based on ether [151,166]. Research in the U.S. had shown that a better (and safer) approach was to use tributyl phosphate (TBP) in kerosene, and following commercial demonstration in Ohio in 1954, it was adopted at Eldorado in 1955.

A simplified process diagram for uranium ore milling at Eldorado's Port Hope, Ontario refinery is given in Figure 5.10.

```
┌──────────────┐     ┌─────────────────────────────┐
│ Uranium      │────▶│ Digestion, in nitric acid   │
│ yellowcake   │     │                             │
└──────────────┘     └─────────────────────────────┘
                                   │
                                   ▼
        ┌─────────────────────────────┐      ┌──────────────────────────────┐
        │ Solvent extraction, with    │─────▶│ Raffinate, of residual salts │
        │ tributyl phosphate in       │      │                              │
        │ kerosene                    │      └──────────────────────────────┘
        └─────────────────────────────┘                    │
                       │                                    ▼
                       ▼                      ┌──────────────────────────────┐
        ┌─────────────────────────────┐      │ Recovery, of nitric acid for │
        │ Extract treatment, twice, by│      │ recycle                      │
        │ solvent extraction with water│     └──────────────────────────────┘
        │ to produce UO₂(NO₃)₂6H₂O    │
        └─────────────────────────────┘
                       │
                       ▼
        ┌─────────────────────────────┐
        │ Boildown, TBP removal then  │
        │ concentration in evaporators│
        └─────────────────────────────┘
                       │
                       ▼
        ┌─────────────────────────────┐
        │ Denitration, by thermal     │
        │ decomposition to UO₃        │
        └─────────────────────────────┘
                       │
                       ▼
        ┌─────────────────────────────┐
        │ Packaging, of UO₃ in ~100 L │
        │ drums                       │
        └─────────────────────────────┘
```

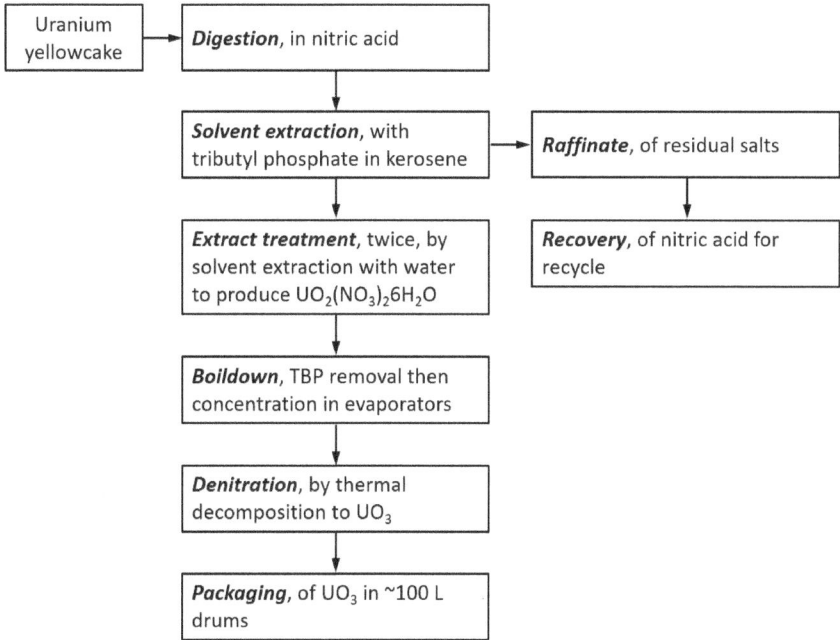

Figure 5.10. Simplified process diagram for uranium ore refining, *circa*. 1958, at Eldorado's Port Hope, Ontario refinery.

- **Digestion.** Yellowcake was digested in a series of three digestion tanks. In the first tank it was dissolved in nitric acid (55 mass% HNO_3) and then cascaded through the next two tanks. Along the way the process temperature was reduced to 35 °C and the free acid concentration to about 3 molar. A vent system was used to draw off the hot, acidic gases from which nitric acid was recovered and recycled back to the digestion tanks.
- **Solvent extraction.** The aqueous "digest-liquor" was passed to a liquid/liquid extraction column packed with perforated stainless-steel trays (to produce mixing) and pumped in from the top of the column, using a pulse-pump. The organic phase, comprising kerosene containing tributyl phosphate (25 volume% TBP), was introduced into the bottom of the column, also with a pulse-pump, and the uranium was selectively extracted from the aqueous phase into the organic phase as these two phases were mixed, in counter-current fashion.
- **Raffinate treatment.** The impurity-containing aqueous phase was high in free nitric acid, so a separate plant process was used to recover the nitric acid and recycle it back to the digestion tanks.

- **Extract treatment.** The uranium-containing organic phase was concentrated in a settling tank followed by a centrifuge, and then pumped to a "scrub column," which was a second solvent extraction column in which a small amount of water was mixed-in to remove residual impurities and nitric acid. From the scrub column, the organic phased was pumped to the "re-extraction column," which was a third solvent extraction column in which the organic phase was mixed with a large excess of water in order to transfer the uranium back (as uranyl nitrate) to the aqueous phase.

- **Boildown.** The aqueous uranyl nitrate solution was cooled and any entrained TBP was removed by skimming. The concentration of uranyl nitrate was then increased by boiling the solution in large evaporators, yielding a uranyl nitrate hexahydrate product.

- **Denitration.** The uranyl nitrate hexahydrate was then converted to uranium trioxide (UO_3) by thermal decomposition.

- **Packaging.** The uranium trioxide, in granular powder form, was cooled and packed into 100 L (25 gal.) drums for shipping to the U.S. Atomic Energy Commission.

Several other processing options were developed for the Eldorado refinery in the 1950s [8]. By 1950 the refinery was able to produce metal-grade uranium oxide. A "green salt" plant and a metal plant were developed in 1957/58, and Eldorado began producing nuclear-grade uranium metal and metal oxide for the Atomic Energy of Canada Ltd. (AECL) nuclear reactor program. By 1958 the refinery was able to produce nuclear-grade uranium as the metal, as uranium dioxide, (UO_2), or as uranium tetrafluoride (UF_4) [167]. These additional processing and product options are illustrated in Figure 5.11.

- **Uranium trioxide.** This was the result of the process sequence illustrated above and in Figure 5.10, in which uranyl nitrate hexahydrate was heated to yield uranium trioxide (γ-UO_3, orange oxide) by thermal decomposition.

- **Uranium dioxide.** The uranium trioxide was cast into pellets and, in a moving bed process, heated and contacted with hydrogen gas to reduce the uranium to yield uranium dioxide (UO_2, brown oxide).

- **Uranium tetrafluoride.** The uranium dioxide pellets were placed in a separate vessel in which they were heated and contacted with hydrogen fluoride gas to produce uranium tetrafluoride (UF_4, green salt). Excess acid was removed by scrubbing, and the salt was separated out by filtration.

96

- *Uranium metal.* The uranium tetrafluoride was mixed with magnesium metal and heated to 1,900 °C, to reduce the uranium to its metal form.

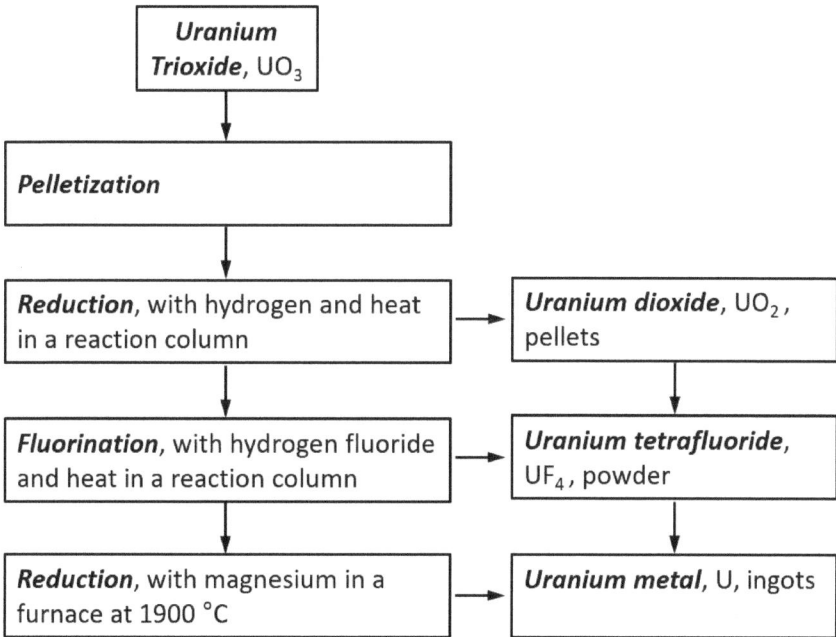

Figure 5.11. Simplified process diagram for uranium products, *circa*. 1958, at Eldorado's Port Hope, Ontario refinery.

The reason for the diversification in products was the AECL CANDU[46] nuclear reactor development program [167], which, in its early developmental stages, used uranium tetrafluoride as the fuel, but later shifted to using uranium dioxide as the fuel. One of the advantages of the Canadian CANDU design was its ability to use natural uranium as the fuel, without the need for enrichment [168]. Enriched uranium was necessary for export to the United States however, and for this purpose either the uranium tetrafluoride or uranium dioxide, would be re-dissolved, mixed with enriched uranium hexafluoride gas (obtained from the United States), and then reprecipitated.

---

[46] CANada Deuterium natural Uranium reactor.

6 CLOSURE AND ABANDONMENT

6.1 Closure.

All mining operations ended in 1959, the Nicholson campsite was abandoned, the workings were flooded, and most of the equipment was removed [137]. Although in subsequent decades there were numerous additional exploration activities in the region of zones No. 1, 2 and 6 [81], the mine itself was never reopened.

At closure in 1959, there were a number of openings and exposures [147]:

- In zone No. 4, the main (#1) shaft was 213 m (700 ft.) deep, with *main levels* at 66 m, 110 m, 152 m, 183 m, and *sublevels* at 35 m, 66 m, and 81 m. There was also one identified raise with the possibility of a second raise (that had not yet been located in the field as of early 2017), plus several surface breakthroughs and some very thin crown pillars that may have failed,
- In zone No. 2, the prospecting shaft was 41 m (135 ft.) deep, with 172 m (562 ft.) of lateral workings completed at the 30 m (100 ft.) level. There was also one adit and some trenching, and
- In zone No. 1, the prospecting shaft was 18 m (60 ft.) deep, with no lateral workings.

Otherwise, the headframe (Figure 6.1) and buildings were left standing but abandoned.

Figure 6.1. The head frame and boiler house in 1981 (Saskatchewan Energy and Resources).

6.2 Post-Closure.

As noted in Chapter 2, following the mine's closure, there were several attempts to find additional economic veins of uranium and/or other metals at the Nicholson site. Although the camp-site was not permanently inhabited in the next three decades after the mine closed, debris left in the surviving structures suggests the camp-site buildings were partially inhabited for periods of time by teams of exploration geologists. This probably involved using some of the buildings for storage, staging, and temporary accommodations.

Between 1989 and 1991, some post-closure steps were taken by Saskatchewan Environment [81,169]:

- Two adits that had been developed in zone No. 4 near the headframe were filled and sealed with waste rock in 1989,
- The headframe from the main mine was removed by intentionally burning it down in September of 1989 (Figure 6.2). Scrap material was pushed into the shaft and the opening covered with a concrete pad in 1991,

Figure 6.2. The Nicholson headframe being burned in 1989 (Photographs courtesy of George Bihun, Saskatchewan Ministry of Environment).

- Shafts number 2 and 3 (see Figure 2.4) were also each covered with concrete slabs in 1989, and
- Two adits were closed by grading rock into and in front of them. A third adit at the end of a trench near shaft number 2 was difficult to access with machinery, so it was secured with a steel mesh grate in 1991 (Figure 6.3).

Figure 6.3. An upper adit at the Nicholson Mine. The right-hand photo shows the view through the grate that is shown in the left-hand photo. (Saskatchewan Research Council, 2015).

When Clifton Associates Ltd., working for Saskatchewan Environment, visited the Nicholson site in 2001 [81] they found that:

- the concrete caps on shafts #1, 2, and 3 were all still in good to excellent condition, with no sign of subsidence or slumping (Figures 6.4 and 6.5), and
- the adit at the end of a trench near shaft number 2 was missing the steel mesh wire cover that had been installed in 1991 (see Figure 6.6).

There were other exposures as well. For example, Figure 6.7 shows a stope breakthrough that had been covered over with a steel mesh grate at some time in the past.

Figure 6.4. The cap on shaft #1 in 2001 (Saskatchewan Environment, [81]).

Figure 6.5. The cap on shaft #2 in 2001 (Saskatchewan Environment, [81]).

Figure 6.6. Trench leading to an adit in 2001 (Saskatchewan Environment [81]).

Figure 6.7. A covered stope breakthrough at the Nicholson Mine (Saskatchewan Research Council, 2015).

Although the headframe had been taken down by Saskatchewan Environment in 1989, by the early 2000s many features of the original mine and campsite infrastructure remained in place, and were continuing to deteriorate, including:

- Numerous mine buildings, including the boiler house, power house, electrical sub-station, auxiliary power house, water tower, mine office, blacksmith and carpenter/machine shops, and warehouse (see, for example, Figures 4.6, 4.7, 4.9, 4.10, 4.11, 6.8),
- Mine equipment, including both mine-shaft cages, the trammer chassis, and many of the tram cars and other hardware (see, for example, Figures 3.5, 3.7, 3.8), and
- Campsite buildings, including the bunkhouse and at least four other residence buildings, the cookhouse, and the auxiliary power house, among others (see, for example, Figures 4.16, 6.9, 6.10).

Figure 6.9 illustrates some of the deterioration of the buildings over time, as the cookhouse roof completely collapsed between 2001 and 2015. In some cases, buildings had almost completely collapsed by 2015, as shown in Figure 6.10.

Figure 6.8. Inside the generator building in 2001 (Saskatchewan Environment [81]).

Figure 6.9. The cookhouse in 2001 (top, Saskatchewan Environment, [81]) and again in 2015 (bottom, Saskatchewan Research Council). Note that the roof had completely collapsed in the intervening years.

Figure 6.10. Examples of badly degraded structures in 2015 (Saskatchewan Research Council).

6.3 Hazards.

In the 1950s it was generally believed that uranium mining and milling posed no particular radiation hazards as long as adequate ventilation was provided, although the need to control dusts was recognized [24]. At the time, both uranium mining companies and health authorities were aware that radium posed a health risk, however information was lacking on specific exposure risks [24]. It wasn't until 1960 that Canadian regulations covering both uranium and radium exposure were brought in by AECB [24].

In the 1950s and 60s, uranium mines in Saskatchewan were not subject to significant pollution control regulations and mine decommissioning was not required [46]. As the 1960s unfolded, public awareness and concerns about radiation and acidity from abandoned and unremediated uranium mines in Canada began to emerge with regard to the Bancroft and Elliot Lake area mine abandonments [24]. The U.S. Government brought in regulations aimed at reducing uranium workers' exposure to radon in 1966 [22].

In subsequent decades the radiation hazards became much better understood, and public and government expectations for mine decommissioning in general changed substantially. As a result, expectations that abandoned mine buildings be dismantled and removed, and that tailings and waste rock be treated, as necessary, to protect humans and the environment would emerge in later years.

Prior to SRC's remediation of the site several assessments were made of the physical, chemical, and radiological hazards, and of the risks these hazards presented to human and ecological health [34,81,169]. The principal Nicholson site hazards identified as of 2017 are summarized below:

- Shaft openings fenced but caution required,
- Raises and stope breakthroughs possible in Zone 4. Some of these have been fenced, others may be present and un-fenced,
- Open adits (possible gas emissions),
- Thin ground over the underground in Zone 4 (thin pillars),
- Transformers and mine machinery still on site,
- Deteriorating and unstable buildings remaining on site (Figure 6.11),
- Asbestos debris in and around buildings (Figure 6.12), and
- Waste rock piles.

Figure 6.11. A residence building in 2015 (Saskatchewan Research Council).

Figure 6.12. Examples of exposed asbestos on floors (left) and in ceilings (right) in 2015 (Saskatchewan Research Council).

Structural Hazards. In a visit to the Nicholson site in 2015, SRC staff noted the hazardous state of many of the buildings. Many of the mine and campsite buildings had become structurally unsafe, with deteriorating roofing, entrances, stairways, access-ways and even some floors ranging from cluttered to visibly unsafe. Leaking roofs had caused partial or complete ceiling collapses in some of the structures. Even where structures had previously collapsed or been demolished, most of the floors and large amounts of structural debris remained in place. Some examples are shown in Figures 4.10, 4.11, 6.8, 6.9, 6.10, 6.12, and 6.13.

Figure 6.13. Example of a Nicholson mine building with deteriorating walls, ceiling, and flooring, as it appeared in 2015 (Saskatchewan Research Council).

Radiation Hazards. Uranium mines and mills involve some unusual hazards in that they all have the potential for environmental impacts due to radiological toxicity. The half-lives of the radioactive contaminants of primary concern are:

- uranium-238 at 4.5×10^9 years,
- thorium-234 at 24 days,
- thorium-230 at 7.5×10^4 years,
- radon-222 at 3.8 days, and
- radium-226 at 1.6×10^3 years.

In 2000 the Saskatchewan Government launched an assessment of its northern abandoned uranium mine sites in order to prioritize them based on public safety and environmental concerns [34]. Several other studies were conducted in the early 2000s [81,152].

The main radiation hazard to people was assessed as being direct exposure to gamma radiation, due to close proximity to radioactive materials in the three shaft openings, two raise openings, two open adits, two stope break-throughs to surface, and/or the waste rock piles.

Generally, the site exhibited only relatively low gamma radiation levels when surveyed in 2001 [81]. A more detailed, pre-remediation, gamma radiation survey was performed by SRC on July 5, 2014, which found that the gamma radiation levels generally varied from 0.03 μSv/h to 18.26 μSv/h with an average of 0.7 μSv/h.

The waste rock piles were found to have gamma radiation levels ranging from about 0.1 to 1.7 μSv/h (at 1 metre) in 2001 [81]. Regarding the potential for acid rock drainage and metal-leaching, the sulphur contents of the waste rock samples suggest that there is some acid generation potential for these materials, and the copper content of the waste rock may pose additional ecological and environmental concerns.

Asbestos Hazards. Some of the buildings contained asbestos panels [81]. At the time of writing, the total amount and nature of these remain to be determined. In other abandoned mines and mills of the same vintage, however, asbestos has been found in a wide array of forms and locations, including corrugated asbestos roofing boards, 'transite type" asbestos-shingle siding made of a cemented asbestos containing material (ACM)[47], asbestos-board interior-wall sheeting, spray-on asbestos ceiling insulation, asbestos-insulation containing construction blocks, and asbestos-lined hot water heaters, boilers, and pipes [68]. According to the Gunnar Mines company, for example, Limpet asbestos fibre was *"highly regarded ... for its insulating and fireproofing qualities"* [66]. As a result, and because it could be spray-applied, it was widely used in building construction, in the Nicolson mine era. The visible asbestos insulation in buildings at the Nicholson site was generally in very poor condition and found as litter on and around the floors of several buildings (see Figure 6.12).

Other Chemical Hazards. Some of the mine buildings contained quantities of chemicals that had been spilled in various locations, such as from spilled and degraded batteries from underground miners' lamps.

---

[47] Transite-type ACM products typically contained about 10-50% asbestos fibre, were common in the 1930s through 1980s, and were used in many applications in which a fire retardant was needed.

Fluorescent light fixtures were common in the more modern of the buildings and some or all of their ballasts probably contain polychlorinated biphenyls (PCBs), although not in very large quantities. There remain, however a number of large transformers outside the electrical substation (Figure 4.8).

Some of the barrels left in various locations around the site were filled, or partially-filled with unknown materials.

6.4 Status in the Early 2000s.

Very little had changed at the Nicholson site by the early 2000s, except for continued deterioration of the site. As already noted, an inspection by Clifton Associates in 2001 [81] found that the concrete pads covering the three shafts were still in good condition, but the steel-mesh wire covering from the adit at the end of a trench near shaft number 2 was missing. Inspection also identified:

- Numerous (15) buildings in various states of disrepair including some that had partially or completely collapsed,
- Metal debris, including drill rods, pipes, barrels, wire, bolts, and steel and electrical cables,
- Equipment, including electric motors, and large electrical transformers, some of which were leaking oil, and
- A core storage area.

Previous work commissioned by the province had assessed all of the significant abandoned mines in Northern Saskatchewan, including uranium mines [81]. Considering both public safety and environmental criteria, the Nicholson site was ranked in 2002 as being #7 of the 22 abandoned mines that had been assessed, with the identified risks being more in terms of public safety than environmental [81].

The Nicholson Mine situation was not uncommon for uranium mines. Worldwide, only a handful of uranium mines have been completely or substantially remediated, including the original Shinkolobwe mine, Democratic Republic of the Congo, in the early 2000s [170]. According to the International Atomic Energy Agency, many parts of the world have experienced large delays in advancing the decommissioning and remediation of nuclear sites, and for a variety of reasons[48] [171]. Only a few of Canada's uranium mines have been completely or substantially remediated:

- The Cluff Lake mine in Saskatchewan was remediated as of 2013 by AREVA Resources Canada Inc. [172],
- The Agnew Lake mine in Ontario was remediated as of the early 1990s by Kerr Addison Mines [172],

---

[48] Including issues related to national policies and frameworks (or the absence thereof), financing, availability of technology and/or infrastructure, stakeholders, and/or politics – see reference [171].

- The mines in the Bancroft area of Ontario (Dyno, Bicroft and Madawaska) were remediated in the 1980s and 1990s [172],
- Of the 12 mines in the Elliot Lake - Blind River area of Ontario, five of the sites had been decommissioned by about 2002, and all of the rest have been decommissioned since that time. At the present time, all of these mine sites have been remediated, with their mine features capped or blocked, facility structures demolished, and the sites landscaped and revegetated [172,173],
- The Rayrock mine in the Northwest Territories was remediated by the Government of Canada in 1996 [172],
- The Port Radium mine, also in the Northwest Territories, was partially decommissioned in 1984 and fully remediated by the Government of Canada by 2009 [172,174],
- In northern Saskatchewan, the Lorado Uranium Mine operated between 1957 and 1960, although its mill continued to operate until 1961. It appears that at some point after its closure, the mine shaft, escape raise, and ventilation raise were all cement-sealed, the head-frame burned-down, and some other remediation conducted. The bunkhouses and some other mine infrastructure were demolished, buried, and the affected landscape recontoured by International Mogul Mines in 1982 [175]. Conwest Exploration Company Ltd. had the mill buildings demolished, burying the debris, and recontouring the "entire mill site area" with clean gravel, in 1990 [176]. This left the tailings, some small previously overlooked legacy infrastructure, a number of small industrial debris sites surrounding the mill area, and the contaminated Nero Lake. All of these were remediated by the Saskatchewan Research Council by 2017 [177,178],
- The Gunnar mine, plus some 35 smaller "satellite" mines in Northern Saskatchewan are currently being remediated by the Saskatchewan Research Council (SRC) [178], and
- The Beaverlodge mine in Saskatchewan comprises 62 licensed properties of which some have been fully remediated while others are still being remediated by Cameco Inc. on behalf of Canada Eldor.

Orphaned or abandoned mines are of particular concern because they represent closed mines whose owner no longer exists, can't be located, or is unable to carry out remediation. In Canada, responsibility for the remediation of such mines reverts to government ("the Crown"). It has been estimated that there are about 10,000 such mine sites in Canada [179].

7 REMEDIATION: THE ENDING

7.1 The Modern-Era Need for Decommissioning and Remediation.

While the end of the Cold War changed the nature and extent of nuclear developments worldwide, this and the Chernobyl accident of 1986 changed attitudes towards nuclear safety and environmental protection. As a result, many nuclear facilities established since the 1950s became redundant, many others reached the end of their design lives, leaving behind large areas of contaminated facilities and land [180]. According to the International Atomic Energy Agency (IAEA), "Many countries were therefore left with facilities requiring to be decommissioned and/or sites requiring to be remediated" [180][49]. Canada is no exception.

The Nicholson site illustrates many of the wide range of adverse impacts of abandoned, orphaned mines, which can include:

- altered landscape, introducing hazards and the loss of otherwise productive land,
- altered vegetation, and the loss, or at least reduced quality of vegetation,
- altered groundwater, introducing hazards and the loss of otherwise potentially productive water,
- altered water bodies and their sediments, introducing hazards and the loss, or at least reduced quality, of the water and/or its constituent plants and animals (such as fish),
- air pollution from dust and/or hazardous gases, and

---

[49] See the glossary for definitions of the terms "*decommissioning*" and "*environmental remediation*."

- physical hazards, including standing structures, pits and shafts, chemical piles and/or tailings dumps.

In cases where abandoned mines have left large amounts of tailings and/or waste rock deposited into unsuitable places, then the sheer volume and mass of material involved mean that it is usually cost-prohibitive to move them to a more suitable location. This is particularly the case for logistically isolated sites such as Nicholson. In such cases the only practical option may be to allow them to remain in place while conducting some remediation to minimize ongoing harm to human health and the environment [181]. The waste rock element of these issues is also applicable to the Nicholson site.

7.2 The Remediation Process Begins.

Canadian federal regulations did not cover uranium mine closures or remediation until May 31, 2000 when the Nuclear Safety and Control Act (NSCA) replaced the Atomic Energy Control Act (AECA). The new legislation was constructed to regulate the complete life-cycle of nuclear activities. As a result, sites like Saskatchewan's Gunnar mine and mill site, that previously existed outside of the jurisdiction of the AECA did come under the jurisdiction of the NSCA and meant that the Gunnar site had to be licensed by CNSC [68,182]. CNSC created a Contaminated Lands Evaluation and Assessment Network Program to identify such sites, evaluate them for safety, and make recommendations for the regulatory approach to each site.

In 2006, the Governments of Saskatchewan and Canada signed a Memorandum of Agreement (MOA)[50] to proceed with the decommissioning and reclamation of 37 Cold War legacy uranium mine and mill sites in Northern Saskatchewan [183,184], which included the remediation of the large Gunnar [68] and Lorado sites, and the Consolidated Nicholson mine and camp-site, among others. As the property owner, the Government of Saskatchewan had primary operational and legal responsibility for the project. The Saskatchewan Research Council (SRC), a provincial Crown corporation, was contracted as project manager and designated agent to manage and perform the required environmental assessment requirements and rehabilitation activities [183].

The Canadian environmental regulatory regime is complex with both the federal and provincial government legislative frameworks applying. The federal government has authority under federal environmental assessment legislation, fisheries legislation, navigable waters legislation and

---

[50] Some of the historical background to the MOA can be found in reference [183].

environmental legislation. The responsible federal departments' oversight is coordinated through the Canadian Environmental Assessment Agency but each department has distinct regulatory applications and authorities.

Specific to the uranium industry, the federal regulatory framework is made more complex with the CNSC being the lead regulatory and licensing authority due to the presence of "nuclear substances" onsite. Due to the nature of the smaller abandoned uranium mines in Northern Saskatchewan, these mines did not have mills and therefore did not have tailings (uranium by-products) stored on their sites. Instead their mined uranium ore was sent offsite for milling. As a result, the smaller abandoned mines did not generally require formal licensing from CNSC[51], and are regulated only through the provincial regulators of Saskatchewan. This has been the case with the Nicholson mine site.

In Saskatchewan, the provincial government is the owner and manager of the Crown Land on which the mine is located and designated permits are required to conduct any work on Crown Land. The Saskatchewan provincial government also has approval requirements under provincial environmental assessment (EA) and environmental protection legislation. The provincial government has a joint agreement with the federal government that allows a coordinated provincial/federal approach to environmental assessment.

The smaller abandoned uranium mines in Northern Saskatchewan have not generally required formal EA approval from either federal or provincial governments as they are too small, from an impact perspective, to trigger the process[52]. In such cases, the remediation work is proposed and approved by Saskatchewan Environment via a permitting process. This is expected to be the case with the Nicholson mine site.

7.3 Community Engagement.

Numerous local, regional and provincial stakeholders were and are interested in the remediation plans and activities, for all of the nearby orphaned and abandoned uranium mines, including the Nicholson site. These include the residents of the closest neighbouring communities of Uranium City and Camsell Portage (with a population of about 120), but also a much broader range of stakeholders. Given the Nicholson site's location in a remote northern area utilized by First Nations, Métis and Northern residents, communications with these local residents has been of paramount importance and is a high priority for SRC throughout its remediation activities. All of the major short-term and long-term site-safety

---

[51] The Gunnar mine, mill, and town-site is a notable exception [68].

[52] Here again, the Gunnar mine, mill, and town-site is a notable exception [68].

and remedial activities for these sites proposed by SRC include consulting with the local population (see for example [152,185-187].

SRC also commissioned a traditional knowledge and traditional land-use study so that the environmental assessment and subsequent remediation could be planned and undertaken in the context of traditional uses of the area. This study was conducted by the Prince Albert Grand Council. Similarly, a socio-economic assessment was conducted by SRC on the potential to use biochar as a soil amendment during land reclamation and revegetation, as a possible way to achieve mutually beneficial outcomes for northern communities, as well as the remediation project [188].

A Project Review Committee (PRC) was formed in the early stages to provide a forum that would ensure involvement of each of the impacted communities and enable them to provide direct input on desired remediation endpoints and options, as well as advice on opportunities to maximize the involvement of northern residents in the economic activities emanating from the project. The PRC was established with the assistance of the Prince Albert Grand Council (PAGC) and included elected officials from: Prince Albert Grand Council, Fond du Lac First Nation, Black Lake First Nation, Hatchet Lake First Nation, Settlement of Uranium City, Settlement of Camsell Portage, and Hamlet of Stony Rapids. The first meeting with local Chiefs and Mayors was held in conjunction with a broader town-hall meeting at the Ben McIntyre School in Uranium City in March of 2007, and guidelines for the PRC were finalized and signed by all parties at a meeting in Stoney Rapids in May of 2008[53].

Communications were also established early with the Northern Saskatchewan Environmental Quality Committee (NSEQC). This committee comprises representatives from the northern municipal and First Nation communities that are impacted by northern mining operations in the province, and in particular with the Athabasca Sub-Committee of the NSEQC. The NSEQC monitors uranium mining in Northern Saskatchewan to confirm environmental protection measures and ensure operations are conducted to increase the socio-economic benefits of the surrounding communities. It also serves as a vehicle to enable northerners to learn more about uranium mining activities and to see first-hand the environmental protection measures being employed, as well as the socio-economic benefits being gained [189].

Beginning in 2005, SRC held annual public meetings in Uranium City, including discussion of the activities being undertaken at all of the sites.

---

[53] The PRC operated for many years and was later wound-up at the request of the committee and replaced with additional, broader, community meetings. The committee had succeeded, credibility, trust, and relationships had been established, and the continuing community meetings have been very effective for all concerned.

These public consultations included representatives of the NSEQC, the CNSC and Saskatchewan Environment. Other similar consultations have been held periodically in neighbouring Athabasca-basin communities. Tours of the various sites were also provided periodically for members of the PRC and the NSEQC. This range of engagement provided many opportunities for community information exchanges, discussions, and feedback.

Some of the most common questions raised by local community members were:

- What are the impacts of the project?
- What are the remediation options?
- Are there any training opportunities?
- Are there job opportunities?
- How can we actively participate in the remediation?

Other initiatives have included communicating project plans and progress in accessible northern media such as radio, media interviews, flyers, posters, mail-outs, newspapers, and magazines. SRC developed and continues to maintain a *"Project CLEANS"* website (www.saskcleans.ca) [178] for this purpose as well. Information has also been routinely provided to northern media for inclusion in their publications (e.g., *Opportunities North*). A key feature of much of these communications has been translating key project information into the Dene and Cree languages to ensure the information being provided would be broadly accessible [190].

These kinds of engagement activities were extremely well received by the communities and their leadership:

"… SRC is including communities in the writing of the Environmental Impact Statement … that type of inclusion has been missing in the past and it's a refreshing change to see …"

Diane McDonald, Prince Albert Grand Council (2010, [191])

Overall, public support has been very high for the project given that mine sites that had been abandoned for over 40 years are finally being cleaned up [192,193]. The consultation process helped ensure that the locally impacted community would be comfortable with the rehabilitation activities and final state of the mine sites. In many cases advice from the community was adopted into SRC's plans.

An example is the prioritization of the order in which the sites would be remediated. At the first broad-based community meeting in Uranium City in 2007, maps, photographs, and descriptions of each of the 37 sites were presented, posted for viewing, and discussed. Although previous work commissioned by the province had already ranked the sites based on technical public safety and environmental criteria [81], SRC asked residents of the Athabasca Basin communities to provide their own input as well, based on their perceptions of the accessibility and risks to their families and other potential visitors to the sites. With this input, some of the priorities were modified. Although this consultation process did not significantly affect the timing of the Nicholson remediation (for which both community interests and technical risk assessments agreed quite well), it did influence the timing of some of the other sites' remediation and contributed to a sense by the communities that their views and concerns were being heard and were having an impact.

In another example, based on advice from the advisory committee and community meetings, SRC revised its procurement process for contract work on the project and also arranged training programs, all with the aim of promoting bidding from and hiring of northern contractors and northern employees.

One of the most significant challenges has been meeting local expectations for economic benefits given the limited project funds available. Efforts have been dedicated to training local communities and Aboriginal entities in such aspects as the tendering process, safety practices, equipment operation, and so on. The project work was compartmentalized to allow local participation in a variety of tasks including light equipment operation, as well as tenders developed that encourage use of local workforces. Efforts were also made to allocate project funds to hands-on training opportunities for work occurring outside the tendering process.

When the Gunnar mine and mill site remediation project was approaching the 2010/2011 demolition phase, described elsewhere [68], both the PRC and the communities were engaged in discussions of how to maximize local employment during this phase. Community liaison positions were created to coordinate employment and training opportunities for individuals at the community level. Programs were developed and put in place to train local communities and Aboriginal entities in the tendering process, safety practices, etc., to maximize the ability of local companies to bid on work and to maximize the number of northern residents qualified to work on the demolition. Such programs have made a significant difference, not only for the Gunnar site remediation project, but for all the rest that have and will follow.

Key remaining goals for the Nicholson site, and all of the other 36 sites, are to ensure that at the end of these remediation initiatives all stakeholders are satisfied that the sites pose no significant dangers to public health and safety, are not a source of ongoing pollution or instability, and allow for productive use of the land similar to its original (pre-mining era) uses, or at least for acceptable alternative uses.

7.4 Remediation Begins.

Under Project CLEANS, and subject to regulatory approvals, it is anticipated that the physical Nicholson site remediation project will consist of:

- Demolition of existing building, facilities and structures,
- Closure of mine openings,
- Appropriate disposal of materials resulting from demolition,
- Rehabilitation of the existing waste rock piles as required,
- Rehabilitation of additional risk(s) as warranted,
- Site characterization,
- General site, hazardous waste, and historical debris clean-up,
- Re-vegetation of areas of the rehabilitated site as required, and
- Appropriate monitoring during and after rehabilitation.

The Nicholson site was first visited (by SRC) for this purpose in 2013, followed-up by initial planning for its remediation activities. At that point in time, site activity was limited to installation of security fencing around the openings to ensure public safety, with a follow-up trip in 2014 to complete preliminary pre-remediation gamma radiation surveying.

Further site investigation was continued in 2015, with a review of all available historical information. Site mine plans and documents were georeferenced to provide a better understanding of the historical layout of the site and a further understanding of the underground mining that had been conducted. Surface water samples were collected in and surrounding the site, beginning in July 2015, in order to establish a baseline for future water sampling programs.

LiDAR[54] (Light Detection and Ranging) surveying was conducted for the Nicholson site in June 2017. The LiDAR data obtained will assist SRC with understanding the relation of the underground workings to that of the surface expressions on site. This data will be instrumental on providing accurate risk assessment on the underground workings. Additionally, UAV[55] orthophotography was also acquired for the site. This will aid in both current site remediation activities and long-term monitoring.

In May 2017, Golder & Associates began the Phase 1, desktop Underground Risk Assessment of all underground site workings, and a

---

[54] LiDAR is a remote sensing method used to examine and create 3D digital maps of the surface of the earth (see Glossary).
[55] Unoccupied Aerial Vehicle, or "drone."

Phase 1 Environmental Assessment was also started. This assessment also included site visits conducted by Golder in late August 2017. In the fall 2017, SRC has installed additional safety fences around the buildings on site (Figure 7.1), as their dilapidated condition was judged to pose a public safety hazard.

Figure 7.1. Fencing and warning signage, as applied to the blacksmith shop in 2017 (Saskatchewan Research Council).

In August 2017, SRC had repairs made to highway 962, the "Goldfields road," which runs from Uranium City, past the abandoned Lorado Mine, to the abandoned Box (gold) Mine at Goldfields (Figure 7.2). This work was aimed at making the road accessible for truck and trailer traffic with access to Lake Athabasca, and also to enable future years' remediation work at the Nicholson mine site. The road repairs included creating a low-level crossing to replace a culvert, repairing and grading portions of the road, clearing a minimum of vegetation along the road, and repairs to the boat launch at the Lake Athabasca shore. Additionally, an archeological survey of work areas where ground was disturbed during the road repairs was completed by CanNorth archeologists. These repairs now enable safer and more rapid access to the Nicholson site, eliminating the prior need for a 60 km (1.5 hr) route by boat around the Crackingstone peninsula (where the Gunnar site is located).

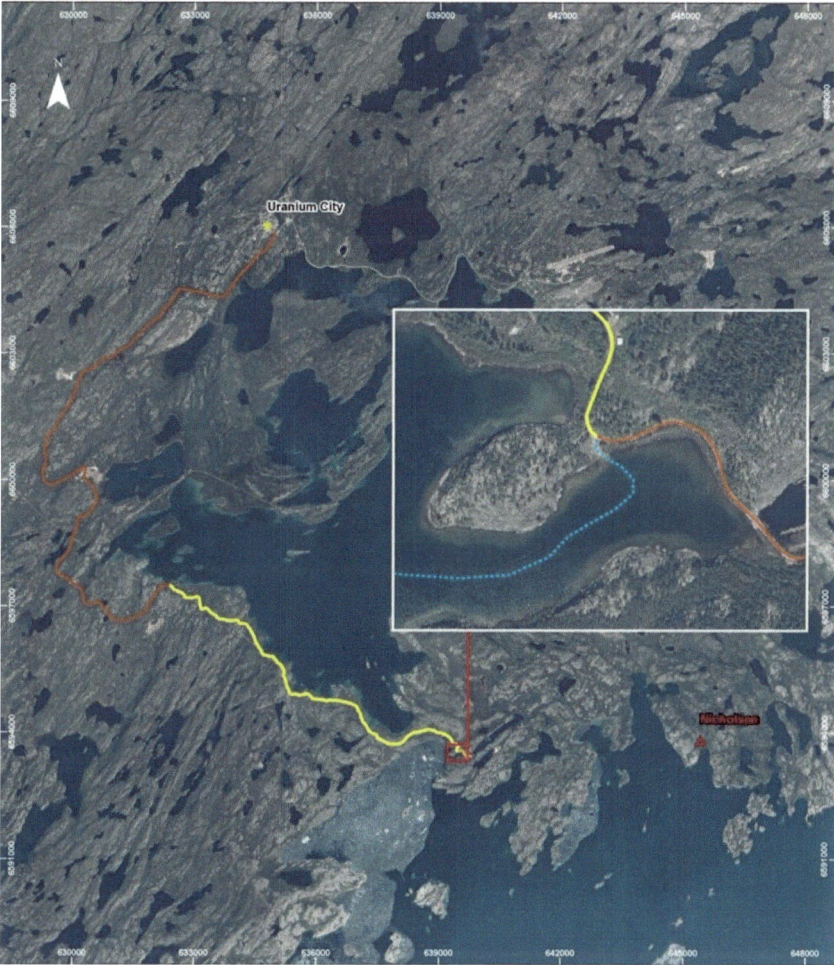

Figure 7.2. Map with topographic background showing the repaired section (solid yellow curve) of highway 962 (solid orange curve), permitting road access from Uranium City to the shore of Lake Athabasca at the boat launch (broken blue curve) near the abandoned Box mine. The Nicholson site is shown in the lower right as a red triangle (Saskatchewan Research Council, 2018).

7.5 Next Steps.

SRC plans to use a "decision-tree" approach to the remediation, which is based on the identification of potentially unacceptable risks to human and ecological health from gamma radiation, contaminants and physical structures. It is expected that such risks will be analyzed in the immediate source area, as well as in the contaminant pathways and final receiving environment.

As already noted, standing structures at the mine and camp sites are degrading, and will need to be demolished. A plan for this, which should include a hazardous materials inventory, demolition strategy, and a site-specific health and safety plan, is being developed. For example, a detailed assessment of the asbestos hazards in the structures will be needed so that precautions can be taken to minimize asbestos exposure to the demolition workers (this was a major issue during the Gunnar mine and mill site facilities demolition in 2010 and 2011 [68]). Given the remote location of the site, the main option available for disposal of the demolished mine and camp-site buildings and other structures will be to landfill the material in one or more landfill sites (in which case suitable locations would have to be prepared and ultimately covered).

Beyond the issue of the standing structures, plans are also being developed for the remediation of the rest the site. Among the issues in this category will be the final disposition of the waste rock piles. Options for remediation of the waste rock include contouring, engineered covers, and/or re-locating some of the piles. The extent to which the covering of any hotspots to ensure the average gamma-ray radiation levels are below 1 $\mu Sv/hr$, plus minor stabilization and re-sloping, where necessary, remain to be determined.

One approach to stabilizing any of the current or future surface cover materials is revegetation. This process involves the use of plant species that facilitate a natural transition towards stable plant communities that reduce wind and water erosion and deep drainage of water. Rather than restoring the original vegetation, in some cases it may be better to introduce non-native species that are better suited to nutrient-lean, acidic soils. In the decades since Nicholson's closure several studies related to the potential for revegetation in the same general region have been conducted [194].

7.6 Transfer to "Institutional Control."

It is SRC's intention to ultimately transfer the entire remediated Nicholson site into the Saskatchewan Government's Institutional Control Program (ICP), which falls under The Reclaimed Industrial Sites Act (2007), and for which the province has Reclaimed Industrial Sites Regulations [183]. The endpoint criteria for the Nicholson remediation project will be developed with this process in mind, with a view to requiring minimal maintenance over the very long term. The projected endpoints are likely to include:

- Radiation levels that do not exhibit in excess of 1 μSv/h above background (averaged over a 100m x 100m surface, or with a maximum spot dose in excess of 2.5 μSv/hr)[56],
- All unsafe buildings demolished and all contaminated materials either buried onsite or removed as dictated by risk,
- All mine openings closed and underground working hazards mitigated,
- Waste rock piles stabilized and adjusted as necessary to reduce environmental and safety hazard risks, and
- Any contaminants that pose an unacceptable environmental risk mitigated by containment onsite or removal.

Such endpoints would allow for traditional uses across the Nicholson site, although some specific areas such as the landfill site(s) might not be available for direct public uses such as camping or seasonal habitation. The endpoints might also preclude the location of any permanent structures such as cabins or cottages in or near the Nicholson site.

---

[56] This would allow for the area to be occupied continuously for up to 42 days before the maximum recommended annual dose level of 1,000 μSv would be reached.

8 RETROSPECTIVE

Like other uranium mines of its time, the Nicholson mine boomed briefly, when the industry was "hot," and collapsed when the bulk of the ore ran out. Such Cold-War-era boom and bust cycles were repeated in Canada, near Uranium City, Saskatchewan [37,43,44,46,68] and Elliot Lake, Ontario [24,173,195], and also in the United States, in such locations as Uravan, Colorado; Moab, Utah; Jeffrey City, Wyoming; and Grants, New Mexico [5].

In the case of Nicholson, like so many of the others, a natural resource was developed and exploited benefitting a local community, a region, and even a country, providing economic and national security benefits. On the other hand, these sites left behind a legacy of environmental disruption and damage, human and animal health risks, along with fearsome clean-up costs.

The Nicholson Mine's total uranium production would have produced about 50 tonnes of U_3O_8 after milling at Eldorado. This would have been worth approximately one million Canadian dollars in 1959, which would be worth about $7.6 million in 2017 Canadian dollars[57]. A portion of the approximately $1 million in uranium sales (in 1959) would have gone to the Consolidated Nicholson Mines, for the mining of the ore, and a portion to Eldorado Nuclear, for the milling of the ore. Eldorado may also have made money on the refining of the yellow cake at its Port Hope, Ontario refinery.

Based on the experiences of the nearby Gunnar Mines operation [68], it seems likely that the total uranium royalties paid to the Saskatchewan and Canadian governments would have been in the order of forty to fifty thousand dollars each (in 1959 dollars).

---

[57] $1 Canadian in 1959 would be worth $8.41 Canadian in 2017 according to "Inflation Calculator," http://inflationcalculator.ca/.

In addition, the principal driving purpose of the uranium exploration wave that led to the development and operations of the Nicholson Mine was to find and develop uranium fuel for strategic military purposes during the cold-war era (see Section 1.4). This was surely a success as the uranium produced at Nicholson would have contributed significantly to the cold-war effort. Taken together, the 16 Beaverlodge area uranium mines contributed substantially to this effort in the Atomic Age and Cold War Eras (Section 1.4). How much uranium from the Nicholson Mine ever found its way into nuclear weapons is either unknown or classified, however most of it would likely have found its way into such weapons, or the reserves for such weapons, or both. Nicholson's contributions to atomic bomb research and development were later applied to peaceful uses, such as nuclear medicine and nuclear power. Whether any, or all, of these uses of uranium amount to a net positive or negative benefit for society overall continues to be publicly debated.

The other Nicholson Mine legacy, of course, is the substantial clean-up effort that is just getting underway as of the writing of this book, and for which the final cost is not yet known with certainty.

Although it is inappropriate to compare the cleanup of the Nicholson site to those of much larger mines, the latter do provide illustrations of how expensive the remediation of such legacy hazards from the past can be. For example:

- In Canada, the nearby Gunnar uranium mine produced over 5 million tonnes of uranium ore, from which the Gunnar mill produced over 8 million kilograms of yellowcake (U_3O_8) between 1955 and 1964. The value of the uranium produced by Gunnar is about $1,100 million (in 2016 Canadian dollars), and the latest public estimate of the total remediation cost is $250 million (also in 2016 Canadian dollars) [196,197]. The remediation itself, is still underway [68].

- In the United States, the Mi Vida uranium mine in Colorado produced 12 million pounds of uranium ore during its operating life, "*enough to make at least eighteen atomic bombs*" [22]. The associated Utex mill remediation project has been estimated at U.S.$400 million [22].

- In Australia, the Rum Jungle was a uranium deposit in the Northern Territory, Australia. Discovered in 1949, a mine and mill were constructed in 1952 that operated from 1953 to 1971. The initial 10-year project produced about 3.2 million pounds of uranium oxide [198]. Upon closure, the Australian government decided not to rehabilitate the mine site, as a result of which acid and metals leached into the nearby East Finniss River for many years. In addition, the abandoned open-pit mine was converted to a lake, which also became contaminated. After mining, the area suffered elevated gamma-ray radiation, alpha-ray emitting radioactive dust, and significant radon concentrations in air. Successive attempts to clean up the Rum Jungle site were made in 1977, 1983, 1990, and again in 2009, spending over A$25.7 million. In 2003, a government survey of the tailings piles at Rum Jungle found that capping which was supposed to help contain this radioactive waste for at least 100 years, had failed in less than 20 years. It has been estimated that the final remediation could cost an additional A$100-200 million. Reference [199].

These stories illustrate that, when not properly planned-for from the beginning, the remediation phase of such industrial development can end-up costing as much or more than the value of the original extracted resource. A key lesson is that mine and mill remediation and reclamation are best considered, planned-for, and budgeted-for at the beginning (before mining ever begins), as part of a comprehensive, full-cycle (sometimes referred-to as "cradle-to-grave") approach to uranium development.

9 GLOSSARY

25
A World War II-era code word for uranium-235. *See* Atomic Code Words.

42-17 grade Z
A World War II-era code word for uranium oxide. *See* Atomic Code Words.

ACM
Asbestos containing material. At Gunnar, the exteriors of most, if not all residences had 'transite type" asbestos-shingle siding made of a cemented ACM.

Adit
A horizontal passage, driven from the surface, for the purpose of mining or dewatering in a mine.

AECB
See Atomic Energy Control Board.

AECL
See Atomic Energy of Canada Ltd.

Atomic Code Words
During the World War II era atomic power research and uranium production were conducted in secrecy. In communications among partners Canada, the U.S., and U.K. code words were used to refer to materials such as uranium oxide (42-17 grade Z), uranium-235 (25), and heavy water (polymer) [24].

Atomic Energy
Control Board (AECB) An entity created in August 1946, under the Atomic Energy Control Act, to control and supervise *"the development, application and use of atomic energy"* [24]. The Board had wide regulatory authority that spanned research, mining, production, transportation, and use of prescribed "atomic substances" [24]. AECB was superseded by the Canadian Nuclear Safety Commission in 2000.

Atomic Energy
of
Canada Ltd. (AECL) A Crown Corporation created in 1952 to assume responsibility for the nuclear research program formerly conducted by the National Research Council of Canada.

Atomic Energy
Worker The 1960 AECB regulations defined for the first time the concept of workers in jobs that could cause them to be exposed to nuclear radiation. AECB also regulated the maximum amounts of radiation to which such a worker could be allowed to become exposed. The AECB regulations also defined the maximum amounts of ionizing radiation to which a member of the general public could be allowed to become exposed, at $1/10^{th}$ of the amount for an atomic energy worker. In modern practice the term has become *"Nuclear Energy Worker (NEW)"* and is defined by the Canadian Nuclear Safety Commission.

Atomic Pile The first nuclear reactor cores contained a "pile" of layers of uranium pellets alternating with graphite bricks.

Black Oxide An impure form of uranium trioxide (which was about 95 percent U_3O_8). *See* Refined Uranium.

Bright Orange
Powder A fairly pure form of uranium trioxide, UO_3. *See* Refined Uranium.

Brown Oxide Uranium dioxide, UO_2. *See* Refined Uranium.

Cage
A cage-like elevator car, suspended from a hoist on steel wire rope and used to transport miners and equipment up and down an underground mine shaft. Also called a Mine Cage. At the Nicholson Mine, the cage compartments were also used to hoist ore and waste rock. For this, tram cars were simply rolled into and out of, one or both of the cages. *See also* Skip.

Canadian Nuclear Safety Commission
(CNSC) Canada's modern-day regulator, which regulates the use of nuclear energy and materials to protect health, safety, security and the environment. CNSC was established in 2000 to replace the former Atomic Energy Control Board.

CANDU
CANada Deuterium natural Uranium reactor. Canada's main commercial nuclear power reactor design, which uses pressurized heavy water as the moderator. The first commercial CANDU reactors were developed in the 1950s and 1960s. The acronym CANDU-PHW is sometimes used to distinguish this design from other experimental CANDU designs.

CANDU-PHW
See CANDU.

CNSC
See Canadian Nuclear Safety Commission.

Cobbing
The process of breaking-up (usually blasted) ore in order to separate ore-grade material from waste rock. Hand-cobbing refers to doing this by hand, usually using a hammer.

Core
Cylindrical pieces of rock of various lengths that are cut and brought to surface through diamond drilling.

Dark Brown Oxide
Uranium dioxide, UO_2. *See* Refined Uranium.

Decommissioning

All technical and administrative actions leading to the release of a facility from regulatory control. This usually includes preliminary characterization, preparation and licensing of the strategy and activities, clean-up, decontamination, and dismantling activities, segregation and packaging of radioactive and non-radioactive wastes, and the final radiological monitoring for release. See [180].

Drift

A mined-out region in an underground mine. These would usually be horizontal and/or parallel to the ore deposits. *See also* Stope.

Dygel

See Forcite.

EBR-I

See Experimental Breeder Reactor I.

Environmental Remediation

Activities aimed at reducing radiation exposure from existing or potential contamination of land areas. This usually includes actions aimed at the contamination itself (by reducing and/or confining the source) and/or at the pathways for human and environmental exposure. See [180].

EQC

See NSEQC.

Experimental Breeder Reactor I

(EBR-I) The United States' first electric-power generating nuclear reactor, which was built in Idaho and started-up in December, 1951.

Extract

See Raffinate.

Forcite

A "gelatin dynamite," comprising 30 to 80% nitroglycerin mixed with cellulose, sodium or potassium nitrate, and a hydrocarbon like tar (to make it waterproof). Dygel (a trademark of ICI Canada) seems to have been another gelatin dynamite formulation.

Geiger-Müller
Meter One of the first commercial hand-held radiation detector/counters. The Geiger-Müller Meter uses an ionization-chamber detector of the same name, enabling it to detect alpha particles, beta particles, and gamma rays. Modern versions are still available today.

Green Salt Uranium tetrafluoride, UF_4. *See* Refined Uranium.

High Grading Selectively mining only the highest grades of material in an orebody.

LiDAR (Light Detection and Ranging or, equally, Light and Radar) LiDAR is a 3D-surveying, remote sensing method used to examine the surface of the earth. In LiDAR, pulsed laser light is used to illuminate a target area, and then the reflected pulses are used to create 3D digital maps of the target area.

Mine Cage *See* Cage.

Mine Skip *See* Skip.

Muck Rock, including both ore and waste rock, that has been blasted from a mine face. Mucking refers to the gathering and transporting broken-up ore and waste rock in a mine.

Mucking *See* Muck.

National Research
Experimental
Reactor (NRX) Canada's second nuclear reactor. It was built at Chalk River, Ontario and commenced operation in 1947. NRX was a 10 MW (later 42 MW) heavy-water-moderated research reactor. It was built and operated by the National Research Council until 1952 and thereafter by Atomic Energy of Canada Ltd. It was closed in 1993. *See also* National Research Universal Reactor and Zero-Energy Experimental Pile Reactor.

**National Research
Universal
Reactor**   (NRU Reactor) Canada's third nuclear reactor. It was built at Chalk River, Ontario and commenced operation in 1957. NRU is a 135 MW heavy-water-moderated research reactor. As one of Canada's national science facilities it is used to generate isotopes for medical diagnoses and/or treatments, to generate neutrons for the Canadian Neutron Beam Centre, and it is also used in the development of CANDU reactor fuels and materials. It is still in operation. *See also* National Research Experimental Reactor and Zero-Energy Experimental Pile Reactor.

NEW   Nuclear Energy Worker. *See* Atomic Energy Worker.

NPD Reactor   *See* Nuclear Power Demonstration Reactor.

NRU Reactor   *See* National Research Universal Reactor.

NRX Reactor   *See* National Research Experimental Reactor.

NSEQC   The Northern Saskatchewan Environmental Quality Committee, comprising representatives from the northern municipal and First Nation communities that are impacted by northern mining operations in Saskatchewan.

**Nuclear Energy
Worker**   (NEW) *See* Atomic Energy Worker.

**Nuclear Power
Demonstration
Reactor**   (NPD Reactor) Canada's first electric-power generating nuclear reactor, which was built in Ontario and started-up in June, 1962.

Orange Oxide   This term normally refers to a blend of red and yellow iron oxides, however in the uranium industry the term has been used to refer to a form of uranium trioxide (γ-UO_3) that is bright orange in colour. The latter is more commonly referred-to as "orange powder." *See also* Refined Uranium, Yellowcake.

Orange Powder   A fairly pure form of uranium trioxide, γ-UO_3. *See* Refined Uranium.

Pachuca   A type of large mineral processing vessel in which ore slurry and air are mixed by countercurrent flow. It is named for Pachuca, a city in Mexico that is famous for its long-standing silver and gold production. The origin of the name may be the Spanish term *pachoacan* (place of silver and gold).

Polymer   A World War II-era code word for heavy water. *See* Atomic Code Words.

Radium Ore   An older term used to refer to the uranium mineral pitchblende. In the 1920s and '30s, uranium minerals were of interest to prospectors as an indicator of radium potential.

Radon   Radon is a chemical element that occurs naturally as a decay product of radium, which in turn is a decay product of uranium. As a result, radium and radon tend to be found wherever there is uranium. Radon poses human health concerns, not so much from radon itself, but from the alpha particles emitted from its decay products: polonium-210 and polonium-214, which can be adsorbed onto fine solid particles and/or small water droplets in the air, then inhaled, and then trapped in the lungs. In this case the alpha particles emitted can directly irradiate lung tissue, which can cause lung cancer [22,24].

Raffinate   In an industrial chemical separation process, where solvent extraction is used to remove components from a liquid, the phase containing the removed material is referred to as the extract, while the remaining liquid from which components have been removed is referred to as the raffinate. Depending upon the process, either phase may contain the desired product. In the case of the solvent extraction of uranium, the raffinate contains the unwanted residual components ("waste").

Raise
A vertical, or nearly vertical, opening in an underground mine that leads from one level to another, and sometimes all the way to the surface.

Refined Uranium
In the refining of uranium any of a number of compounds may be the final, or intermediate, product. In very early refineries, yellowcake was converted into a form of uranium trioxide called "black oxide" (which was about 95 percent U_3O_8). Later refineries produced a more pure form of UO_3, called "orange powder" (or "bright orange powder," or "orange oxide") due to its yellow-orange colour. In some refineries uranium trioxide was reduced to uranium dioxide, UO_2, called "brown oxide" (or "dark brown oxide") due to its dark brown colour, and then converted to uranium tetrafluoride, UF_4, called "green salt," due to its green colour. *See* Figure 5.9. *See also* Yellowcake.

Skip
A bucket-like container, suspended from a hoist on steel wire rope and used to transport mined ore and waste rock up an underground mine shaft to the surface. Also called a Mine Skip. Skips were not used at the Nicholson Mine. Instead, tram cars were simply rolled into and out of, one or both of the cage compartments. *See also* Cage.

Saskatchewan Research Council
(SRC) A research and technology organization incorporated as a Crown Corporation and owned by the Government of Saskatchewan. SRC conducts independent applied, research, development, demonstration, testing, and commercialization.

SRC
See Saskatchewan Research Council.

Stope
In underground mining a stope is the ore surface being mined and/or the open passageway space that is left behind after the ore has been mined. A near-horizontal such passageway is termed a drift. Stoping refers to the removal of the ore from this space, and is practised when the surrounding rock is stable enough not to collapse after the ore has been mined out. *See also* Drift.

Transite   *See* ACM.

Tube Alloys   The code name for the secret atomic weapons development programs of the U.S., UK, and Canada, that were merged in 1943.

U-235   The specific isotope of uranium (U) that is involved in sustainable nuclear fission. U-235 is naturally present only in very low concentrations, less than one percent, in the main uranium isotope, which is U-238. The numbers refer to the relative atomic mass of the element – atoms of U-238 have three more neutrons in them than do atoms of U-235.

UAV   An Unoccupied Aerial Vehicle (sometimes known as a drone).

Vein   A mineral vein is a layer or sheet of crystallized minerals in a rock formation. Such veins would have been created by the precipitation of mineral components from a solution as it flowed through a natural fissure, or crack, in the rock.

Yellowcake   The final precipitated oxides of uranium that result from the milling of raw uranium ore using a leach process. Although often referred to as U_3O_8, for older processes this is a bulk-average approximation. Yellowcake from Cold War-era milling operations was usually a mixture of UO_2 and UO_3 with minor amounts of uranyl hydroxide and uranyl sulphate. The "yellowcake" produced by some mills is (was) actually brown or black, rather than yellow in colour. *See also* Refined Uranium.

ZEEP Reactor *See* Zero-Energy Experimental Pile Reactor.

Zero-Energy
Experimental Pile
Reactor   (ZEEP Reactor) Canada's first nuclear reactor and the
world's first non-U.S. reactor. It was built at Chalk River,
Ontario and commenced operation in 1945. ZEEP was a
heavy-water-moderated reactor and was used to irradiate
uranium to produce plutonium, and also to irradiate
thorium to produce uranium-233. It was closed in 1970.
See also National Research Experimental Reactor and
National Research Universal Reactor.

10 APPENDICES

Appendix 1. **Aerial View of Zones and Buildings at the Nicholson Site (SRC).**

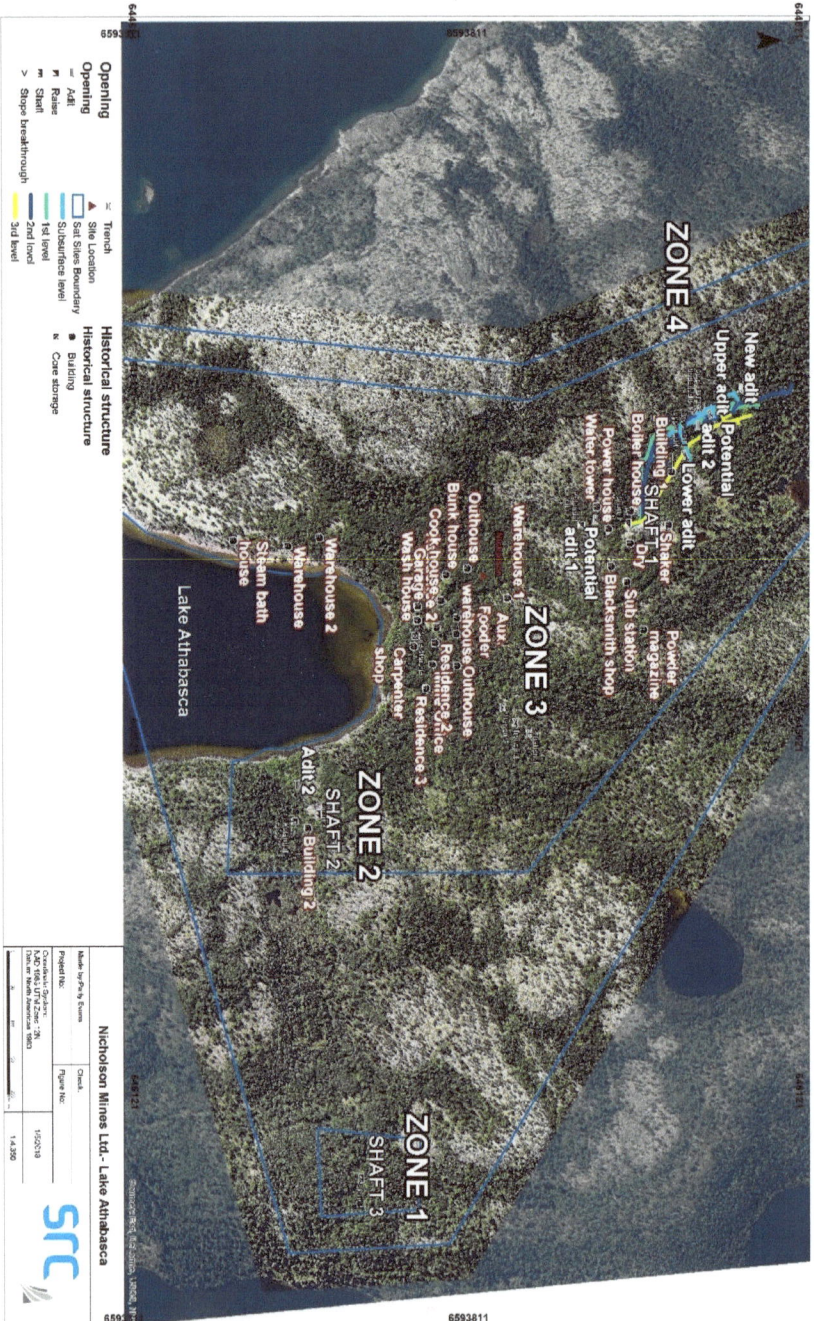

Appendix 2. Expanded View of Zones and Buildings from Appendix 1.

Appendix 3. Topographic View of Zones and Buildings at the Nicholson Site (SRC).

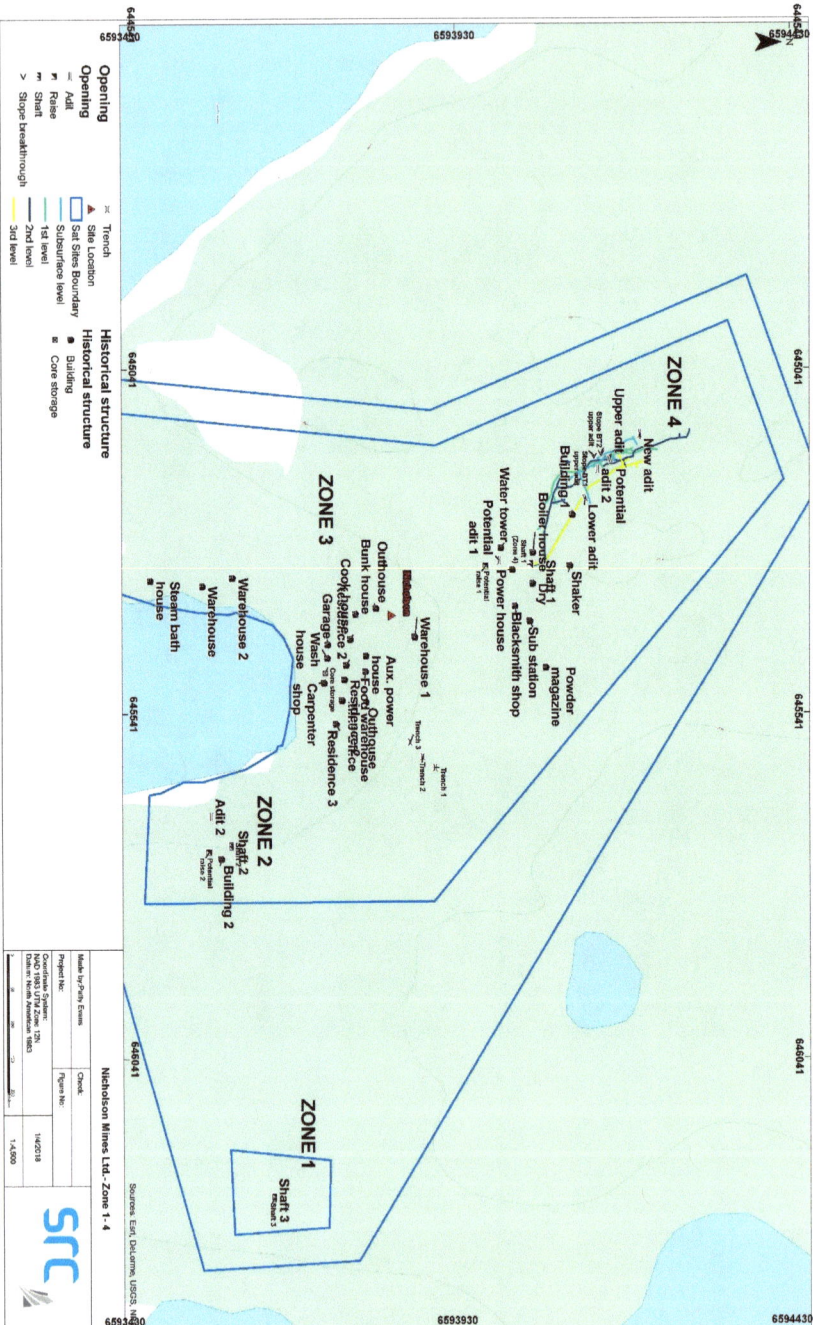

Appendix 4. Expanded View of Zones and Buildings from Appendix 3.

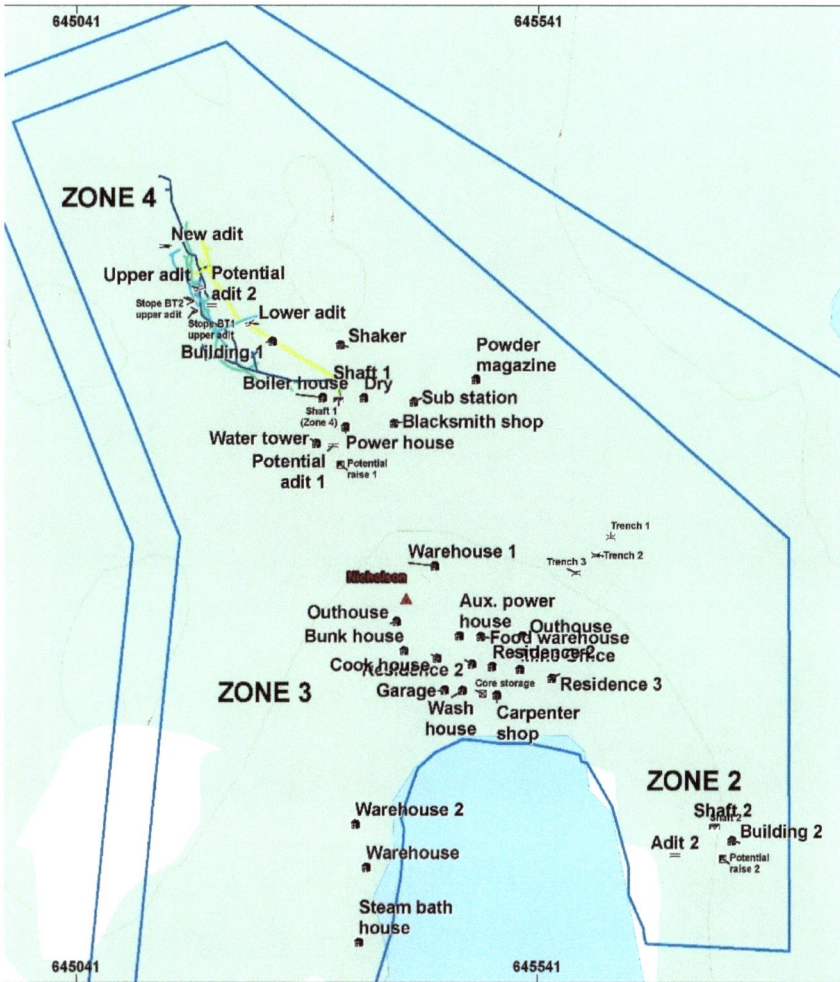

645041   645541

ZONE 4

New adit

Upper adit   Potential adit 2

Stope BT2 upper adit   Lower adit

Stope BT1 upper adit

Building 1   Shaker   Powder magazine

Boiler house   Dry   Shaft 1

Shaft 1 (Zone 4)   Sub station

Blacksmith shop

Water tower   Power house

Potential adit 1   Potential raise 1

Trench 1

Warehouse 1   Trench 3   Trench 2

Nicholson

Outhouse   Aux. power house   Outhouse

Bunk house   Food warehouse

Residence 2

Cook house   Residence 2

ZONE 3   Garage   Core storage   Residence 3

Wash house   Carpenter shop

ZONE 2

Warehouse 2   Shaft 2   Building 2

Adit 2   Potential raise 2

Warehouse

Steam bath house

645041   645541

145

Appendix 5. Layout of the Buildings on the Nicholson Site. (Nicholson Mines Ltd., circa 1950).

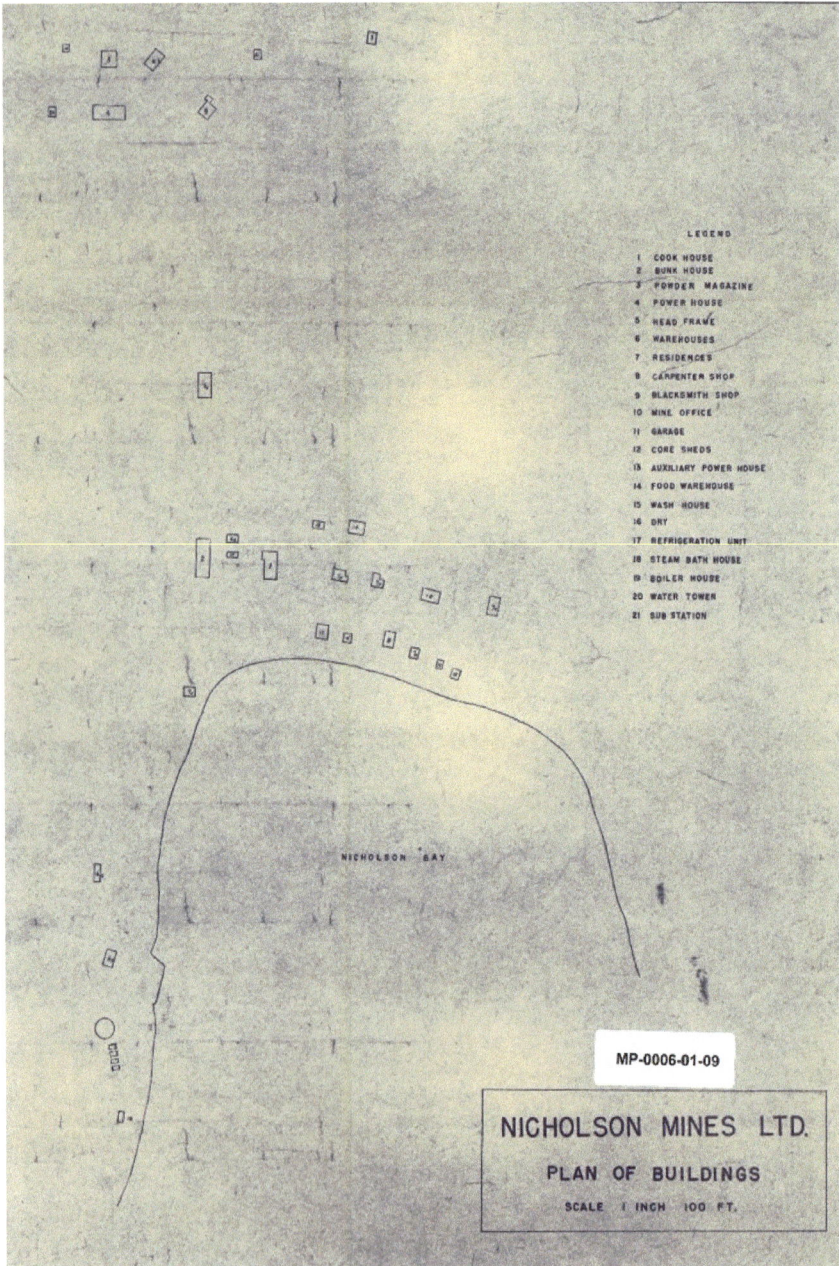

LEGEND

1 COOK HOUSE
2 BUNK HOUSE
3 POWDER MAGAZINE
4 POWER HOUSE
5 HEAD FRAME
6 WAREHOUSES
7 RESIDENCES
8 CARPENTER SHOP
9 BLACKSMITH SHOP
10 MINE OFFICE
11 GARAGE
12 CORE SHEDS
13 AUXILIARY POWER HOUSE
14 FOOD WAREHOUSE
15 WASH HOUSE
16 DRY
17 REFRIGERATION UNIT
18 STEAM BATH HOUSE
19 BOILER HOUSE
20 WATER TOWER
21 SUB STATION

NICHOLSON BAY

MP-0006-01-09

NICHOLSON MINES LTD.

PLAN OF BUILDINGS

SCALE 1 INCH 100 FT.

Appendix 6. Some Police Milestones Relevant to A/Supt. John D. Nicholson and the Nicholson Mine.

Year	Milestone
1873	The Canadian government established the North-West Mounted Police (NWMP) and sent it to police the North-West Territories (the "great North-West" comprised the present-day Alberta and Saskatchewan).
1874	The NWMP established permanent posts in Alberta, at Fort Macleod and Fort Edmonton, and in Saskatchewan, at Fort Pelly.
1875	The NWMP established additional posts in Alberta, at Fort Calgary, and in Saskatchewan, at Fort Walsh.
1876	Another post was established at Battleford, Saskatchewan.
1885	Members of the Force served in the Northwest Rebellion.
1895	The jurisdiction of the NWMP was expanded to include the Yukon, and new posts were established at the summit of the Chilkoot Pass and in Dawson City, Yukon, to deal with the Klondike Gold Rush.
1899-1902	Members of the Force served in the South African War (i.e., the Second Boer War).
1903	The jurisdiction of the NWMP was expanded to include the Arctic Coast.
1904	King Edward VII conferred the title of Royal North-West Mounted Police (RNWMP).
1905-1916	The RNWMP was contracted to conduct provincial policing in Alberta and Saskatchewan. These contracts ended when the RNWMP became short-handed due to the extra duties imposed on them with the outbreak of World War I.
1912	The jurisdiction of the RNWMP was expanded to include northern Manitoba.
1914-1918	Members of the Force served in the First World War.
1917-1928	The Saskatchewan Provincial Police (SPP) force was created. It handled provincial policing until it was disbanded in 1932, after which their duties were assumed by the RCMP.
1917-1932	The Alberta Provincial Police (APP) force was created. It handled provincial policing until it was disbanded in 1932, after which their duties were assumed by the RCMP.
1920	The RNWMP absorbed the Dominion Police and became the Royal Canadian Mounted Police (RCMP), with jurisdiction across all provinces and territories.
1928	The RCMP returned to provincial policing with a new contract in Saskatchewan.
1932-1938	The RCMP took over provincial policing in Alberta, Manitoba, New Brunswick, Nova Scotia and Prince Edward Island.
1939-1945	Members of the Force served in the Second World War.

Appendix 7.   Isometric Drawing of Zone No. 4 Workings (Nicholson Mines Ltd., *circa* 1958). See also Figure 3.9.

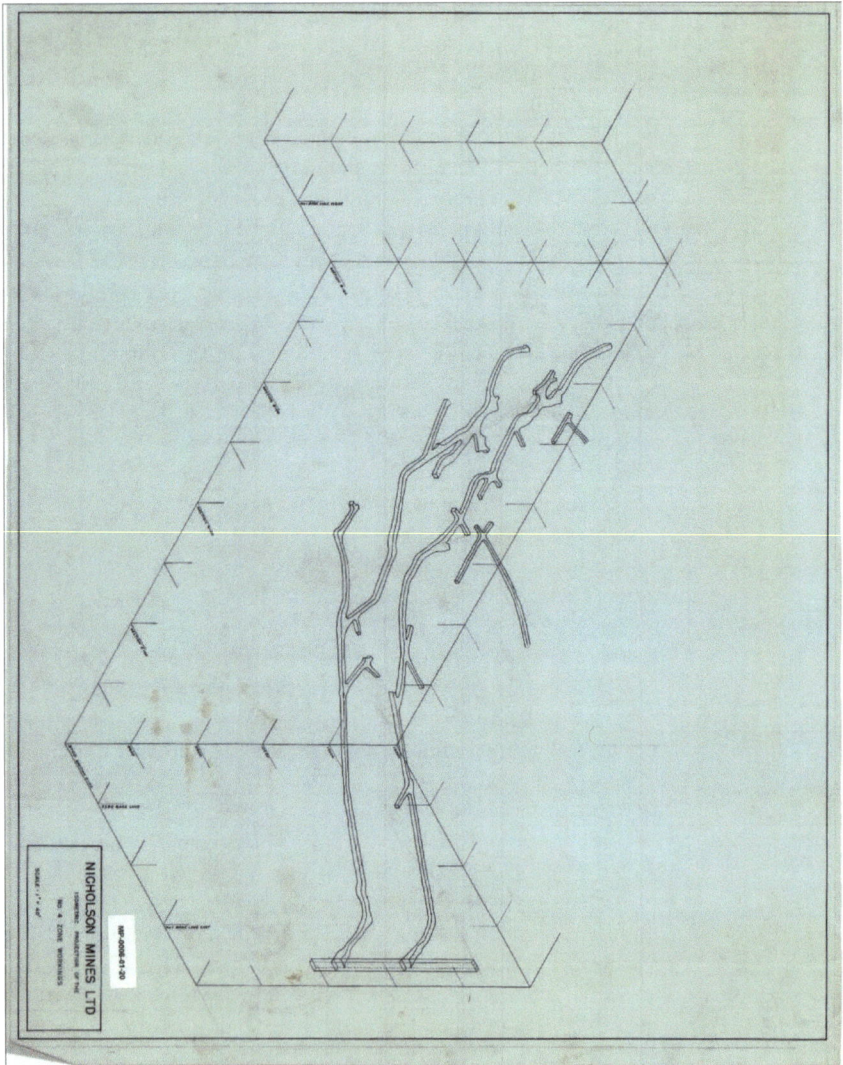

Appendix 8. Section of Zone No. 4. Surface to 100 Level (Northshore Uranium Developers, 1959).

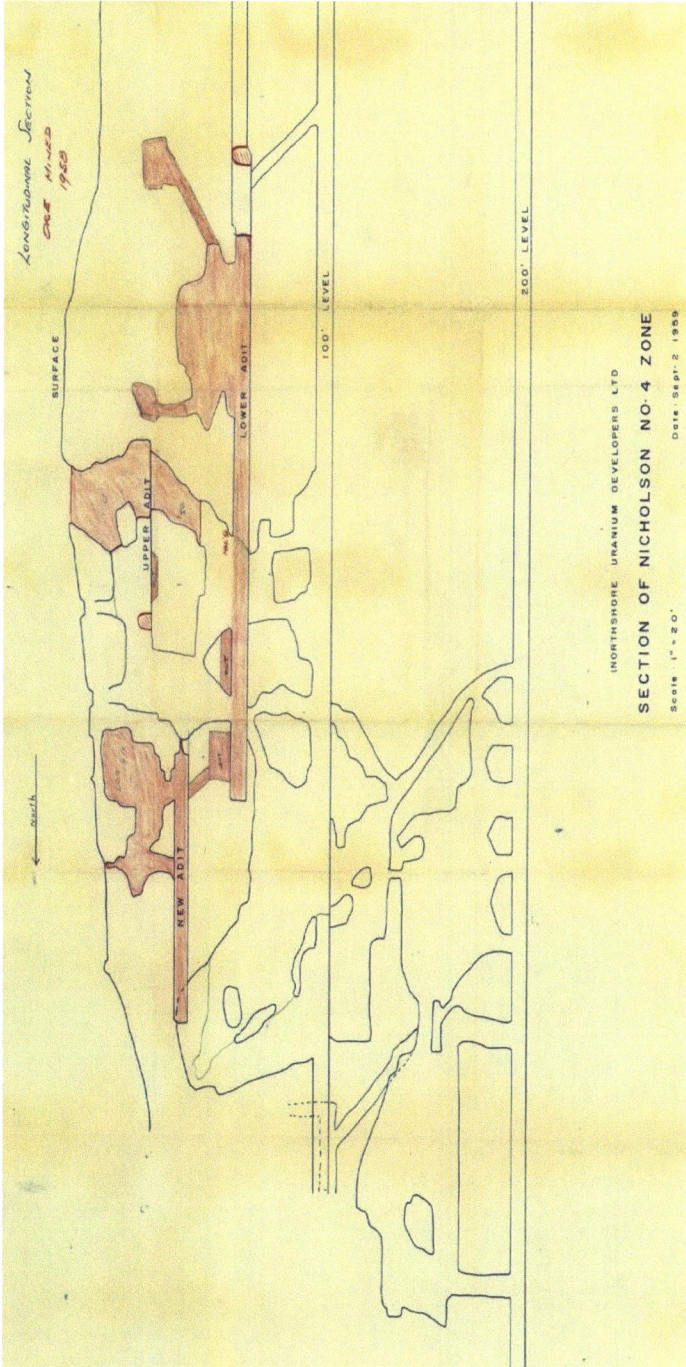

Appendix 9.   Section of Zone No. 4. Surface to 200 Level (Northshore Uranium Developers, circa 1959).

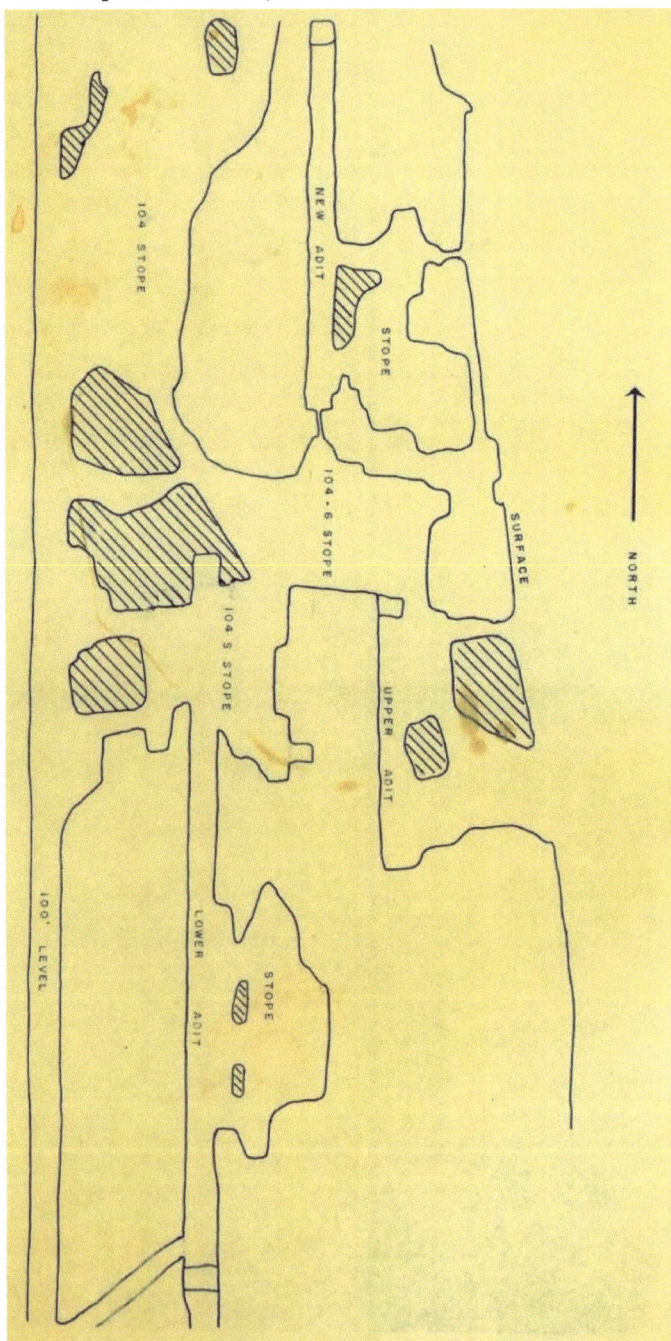

Appendix 10. Zone No. 4 Mine Development as of 1958 (Northshore Uranium Developers, 1959).

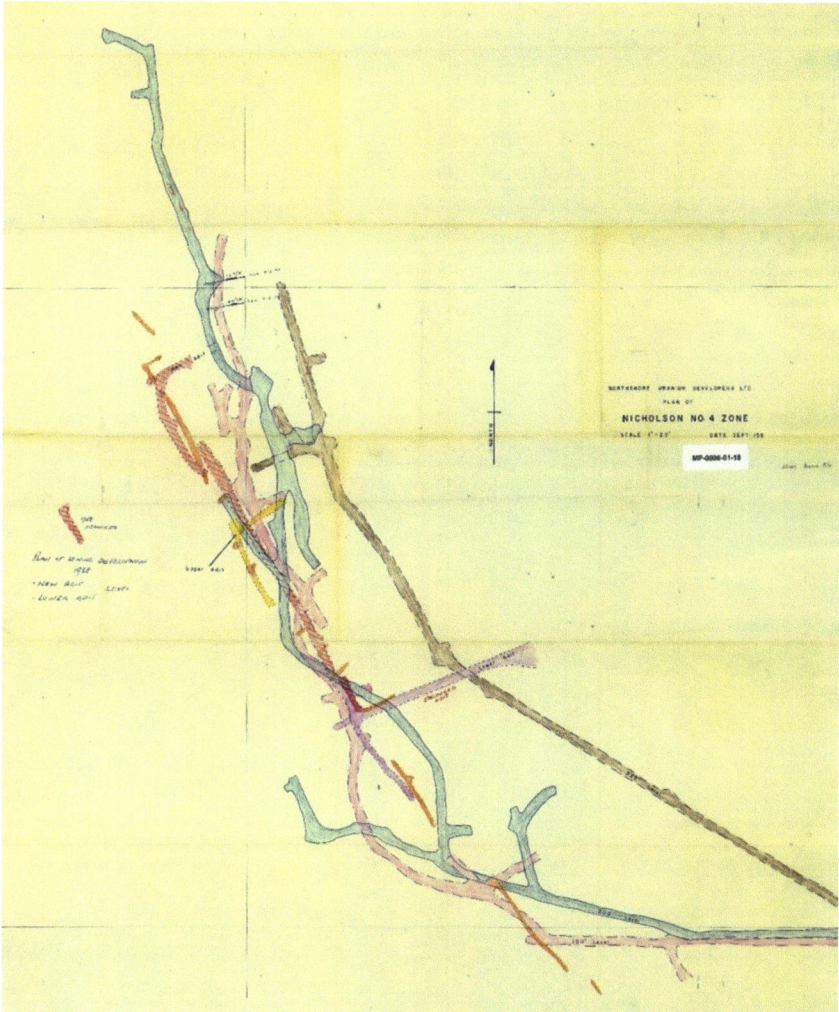

Appendix 11. Approximate Unit Conversions.

These unit conversions are approximate only:

Mass	Imperial pounds to kilograms	1 lb = 0.454 kg
	Imperial tons to metric tonnes	1 ton = 0.907 tonne
Distance	Imperial feet to metric metres	1 ft = 0.3048 m
Volume	U.S. gallons to metric litres	1 US gal = 3.785 l
	Imperial gallons to metric litres	1 Imp gal = 4.546 l

For uranium, 1 tonne of uranium metal in U_3O_8 = 1.1792 tonne as U_3O_8.

11 SUMMARY

The Nicholson Mine,
Saskatchewan's First Cold War Uranium Mine

The first discovery of uranium in Saskatchewan was at Nicholson Bay, in a remote location in northern Saskatchewan, on the shore of Lake Athabasca. Uranium was first noted at what became the Nicholson site in 1929 when uranium was only of interest as an indicator of radium potential. When uranium ores became of strategic national interest in about 1940, a cross-Canada search was launched to find uranium deposits. The first to be found and developed was in the Northwest Territories. The second arose from a return to exploration at the Nicholson site in the Beaverlodge area in 1944. The Nicholson mine was the first uranium mine to be developed in Saskatchewan and, in 1949 was the only active uranium mine in Canada outside of the Northwest Territories. By 1959 the Nicholson ore body had been essentially depleted, but the Nicholson mine had played its role in helping Canada become one of the largest uranium producers in the world. It produced about 12,800 tonnes of uranium ore, yielding about 50 tonnes of uranium (as U_3O_8), and an estimated 60- to 90 thousand m^3 of waste rock. Following closure in 1960, the Nicholson site was abandoned with little remediation and no reclamation being done. Forty-five years would pass before the governments of Saskatchewan and Canada reached an agreement to fund the remediation (clean-up) of the Nicholson site, and contracted the management of the project to the Saskatchewan Research Council (SRC). At the time of writing this book the clean-up was about to begin, with several years of clean-up activity anticipated, and then a period subsequent monitoring activity, before the site is expected to be released into a long-term management and monitoring program.

Print ISBN: 978-0-9958081-4-0
ePub ISBN: 978-0-9958081-5-7

12 ABOUT THE AUTHORS

Dr. Laurier Schramm has over 35 years of R&D experience spanning each of the industry, not-for-profit, university, and government sectors. He is currently President and CEO of the Saskatchewan Research Council (SRC). His interests include technological innovation, management and leadership, colloid & interface science, and nanotechnology. He holds 17 patents, and has published 15 books and over 400 other publications and proprietary reports. He has served on many expert advisory panels and Boards, is co-founder of Innoventures Canada Inc. (I-CAN), and co-founder of Canada's Innovation School™. He has received national scientific and engineering awards for his work, and is a Fellow of the Chemical Institute of Canada and an honourary Member of the Engineering Institute of Canada.

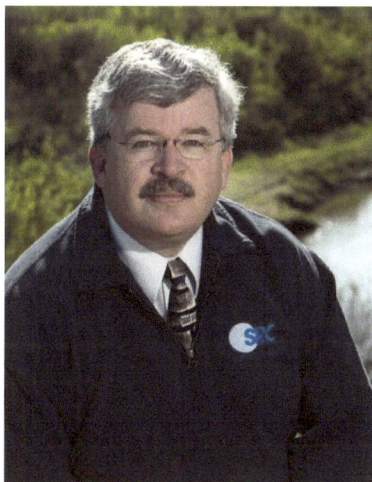

Patty Ogilvie-Evans was born and raised in Saskatchewan, and her interest in earth processes led her to pursue a career in Geological Sciences. She graduated from the University of Saskatchewan in 2006, and has been working as a Geologist in Saskatchewan for the past 12 years. She has a mining background with experience in gold and uranium underground and open-pit mining, diamond exploration and several years of uranium exploration. She is currently a part of Saskatchewan Research Council's (SRC) Environmental Remediation team, working with abandoned legacy sites in Northern Saskatchewan. The history of the abandoned sites is of particular interest to Patty and she greatly enjoys the challenges of finding, locating and recreating abandoned mines to provide a better understanding to assist in the ultimate remediation of these sites.

13 REFERENCES

1. Eldorado, *Uranium in Canada*, Eldorado Mining and Refining Ltd.: Ottawa, 1964.
2. Penrose, R.A.F., *Econ. Geol.*, **1915**, *10*, 161-171.
3. Hahn, O., "From the Natural Transmutations of Uranium to its Artificial Fission," In *Nobel Lectures, Chemistry 1942-1962*, Elsevier, Amsterdam, 1964, pp. 51-66 (Hahn's Nobel Prize Lecture given on 13 Dec. 1946).
4. Ringholz, R.C., *Uranium Frenzy, Boom and Bust on the Colorado Plateau*, Norton: NY, 1989.
5. Amundson, M., *"Yellowcake Towns: Uranium Mining Communities in the American West,"* University Press of Colorado, Boulder, 2004.
6. Davidson, C.F., *The New Scientist*, **1957**, *(Feb. 21)* 9-11.
7. Taft, R.B., *Radium Lost and Found*, Furlong: Charleston, 1938.
8. Griffith, J.W., *The Uranium Industry - Its History, Technology and Prospects*, Mineral Report 12, Dept. of Energy, Mines and Resources: Ottawa, 1967.
9. Guidry, M., *The Guedry-Labine Family and the World's First Atomic Bomb*, accessed December 2013, http://freepages.genealogy.rootsweb.ancestry.com/~guedrylabinefamily /guedrylabineatomicbomb.
10. Bothwell, R., *Eldorado, Canada's National Uranium Company*, University of Toronto Press: Toronto, 1984.
11. Globe and Mail, "Gilbert LaBine: His Tools Were Pick, Paddle and .30-.30," *The Globe and Mail*, **1957**, *July 20*, p. 35.
12. Saskatchewan Department of Mineral Resources, *Inventory and Outlook of Saskatchewan's Mineral Resources*, Report No. 83, Dept. Mineral Resources: Regina, SK, Nov., 1966, 52 pp.
13. Natural Resources Canada, *Atlas of Canada*, 6th Ed., Natural Resources Canada: Ottawa, 2009.
14. Alcock, F.J., "Geology of Lake Athabaska Region, Saskatchewan," Memoir 196, Geological Survey of Canada, Ottawa, 1936.

15. CIM, *The Beaverlodge Uranium District*, Beaverlodge Branch, Canadian Institute of Mining & Metallurgy, Edmonton, Sept., 1957, 57 pp.
16. Saskatchewan Geological Survey, "Geology, and Mineral and Petroleum Resources of Saskatchewan 2003," Saskatchewan Industry and Resources: Regina, Misc. Report 2003-7, 2003.
17. Beck, L.S., "Uranium Deposits of the Athabasca Region," Report 126, Geological Survey, Saskatchewan Mineral Resources: Regina, 1969.
18. La Bine, D.G., *Gilbert A. LaBine 1890 – 1977*, accessed December 2013, http://www.labine.com/gilbert_a_labine, 2004.
19. Hahn, O.; Strassmann, F. "Concerning the Existence of Alkaline Earth Metals Resulting from Neutron Irradiation of Uranium" *Naturwiss.*, **1939**, *27*, 11-15. Translation in *Am. J. Phys.*, **1964**, *January*, 9-15.
20. Meitner, L.; Frisch, O. R. "Disintegration of Uranium by Neutrons: a New Type of Nuclear Reaction," *Nature*, **1939**, *143 (3615)*, 239–240.
21. Peierls, R. "O. R. Frisch, 1904-1979," *Nature*, **1980**, *284 (13 March)*, 196–197.
22. Zoellner, T., *Uranium*, Penguin Books: London, 2009.
23. Rutherford, E., *Radio-Activity*, Cambridge University Press: Cambridge, 1904.
24. Sims, G.H.E., *A History of the Atomic Energy Control Board*, Canadian Government Printing Centre: Ottawa, 1980.
25. Dominion Bureau of Statistics, "Chronological Record of Canadian Mining Events from 1604 to 1943 and Historical Tables of the Mineral Production of Canada," Department of Trade and Commerce, Edmond Cloutier Printer: Ottawa, ON, 1945.
26. CIM Staff, "The Eldorado Enterprise," *Trans. Can. Inst. Min. Met.*, **1946**, *49*, 423-438.
27. World Nuclear Association, *Brief History of Uranium Mining in Canada*, Appendix 1, World Nuclear Association: London, accessed January 2013, http://www.world-nuclear.org/info/Country-Profiles/Countries-A-F/Appendices/Uranium-in-Canada-Appendix-1--Brief-History-of-Uranium-Mining-in-Canada/.
28. Piper, L., *Environment and History*, **2007**, *13*, 155-186.
29. "Early Instrumentation - 1920's," *National Radiation Instrument Catalog 1920 – 1960*, 2007, http://national-radiation-instrument-catalog.com/new_page_144.htm.
30. Taft, R.B., "Radium Hounds," *Scientific American*, **1939**, *160(1)*, 8-47.
31. LaBine, G.A., "Submission to Royal Commission on Canada's Economic Prospects," Government of Canada, Ottawa, 8 March 1956.
32. Hunter, W.D.G., "The Development of the Canadian Uranium Industry: An Experiment in Public Enterprise," *Can. J. Econ. Pol. Sci.*, **1962**, *28(3)*, 329-352.
33. Ross, M.; Hovdebo, D.G., "Uranium Mine Reclamation - A Myriad of Extremes Politics, Perceptions and Long-Lived Radionuclides," in Proc.19th Annual British Columbia Mine Reclamation Symposium, Dawson Creek, BC, pp. 188-196 (1995).

34. Athabasca Interim Advisory Panel, "Athabasca Land Use Plan: Stage One," Saskatchewan Environment: Regina, March 2006.

35. Kneen, J., "Uranium Mining in Canada – Past and Present," Presented to: *Indigenous World Uranium Summit*, Nov. 30-Dec. 1, 2006, Window Rock, Arizona. Accessed at: http://www.miningwatch.ca/sites/www.miningwatch.ca/files/Uranium _Canada_0/

36. Tilman, A., "On the Yellowcake Trail," Parts 1-4, *Watershed Sentinel*, **2009**, *June-July*, 18-22; **2009**, *Sept.-Oct.*, 28-31; **2009**, *Nov.-Dec.*, 28-31; and **2009**, *Mar.-Apr.*, 28-31.

37. Belanger, D.; Hallett, F.; Dusseault, C., *The History of Uranium City*, Self-published. Available through several public libraries including the La Ronge Public Library and the Saskatoon Public Library, 1975, 19 pp.

38. MiningWatch Canada, Elliot Lake Uranium Mines, MiningWatch Canada: Ottawa, 2012, http://www.miningwatch.ca/elliot-lake-uranium-mines.

39. Hutton, E., "The Atom Bomb That Saves Lives," *Maclean's Magazine*, **1952**, *65(4)* February 15, p.7.

40. Fedoruk, S., "The Growth of Nuclear Medicine," *50 Years of Nuclear Fission in Review*, Canadian Nuclear Society: Ottawa, 1989, http://media.cns-snc.ca/history/fifty_years/fedoruk.html.

41. Idaho National Laboratory, "Experimental Breeder Reactor - I (EBR-I)," Brochure 07-GA50535_02, Idaho National Laboratory: Idaho, 2007.

42. Fawcett, R., *Nuclear pursuits: The scientific biography of Wilfred Bennett Lewis*, McGill-Queen's University Press: Montreal, 1994.

43. Grade 10 Class Candu High School, *The History of Uranium City and District*, Lakeland Press, La Ronge, SK, 1982, 63 pp.

44. McBain, L., "Uranium City," Encyclopedia of Saskatchewan, University of Regina: Regina, 2006, http://esask.uregina.ca/entry/uranium_city.html.

45. Nichiporuk, A., "What Does the Future Hold for Uranium City?" *CIM Magazine*, **2007**, (November), 2 pp.

46. Keeling, A., "Born in an atomic test tube. Landscapes of cyclonic development at Uranium City: Saskatchewan," *The Canadian Geographer*, **2010**, *54(2)*, 228-252.

47. Northern Miner, Saskatchewan Uranium Shows Attract Monied Interest," *The Northern Miner*, 1952, Jan. 10, P. 1.

48. "The Uranium Rush - 1949," *National Radiation Instrument Catalog 1920 – 1960*, 2007, http://national-radiation-instrument-catalog.com/new_page_144.htm.

49. US Atomic Energy Commission, *Prospecting For Uranium*, US Government Printing Office: Washington, 1949.

50. Wright, R.J., *Prospecting with a Counter*, U.S. Atomic Energy Commission: Washington, 1954.

51. Northern Miner, "Uranium – Canada Maintains Place in Frantic World Production Race," *The Northern Miner*, 1952, Nov. 27, p.58.

52. Joubin, F.R.; James, D.H., "Canada's Uranium Future," *Precambrian*, **1956**, *29(5)*, 13-14.

53. Maclean's, "Uranium City Here We Come," *Reader's Digest Magazine*, **1954**, *64(384)*, April, 59-64.

54. Richardson, B.T., "The Hottest Square Mile in the World," Maclean's Magazine, **1951**, *64(20) Oct. 15, p. 14.*

55. Life, "Uranium Rush is On in Athabaska," *Life Magazine*, **1952**, *33(7)*, Aug. 18, pp. 15-19.

56. Stapleton, B., "Canada's Great Uranium Rush," *Collier's Magazine*, **1953**, *October 2*, pp. 32-41.

57. Advocate, "Atom Age Mining Rush Begins in N. Canada," *Advocate (Burnie, Tasmania)*, **1952**, *August 5*, p. 3.

58. Courier-Mail , "Began at dawn, Uranium rush in Canada," *The Courier-Mail (Brisbane, Queensland)*, **1952**, *August 5*, p. 1.

59. Sydney Morning Herald, "Canada's First Uranium Rush," *The Sydney Morning Herald (New South Wales)*, **1952**, *August 5*, p. 3.

60. Mercury, "Uranium rush in Canada," *The Mercury (Hobart, Tasmania)*, **1952**, *August 5*, p. 3.

61. TMC, *"The Birth of a Great Uranium Area,"* Documentary Film, Technical Mine Consultants (TMC, Toronto) and Canadian Television Film Production, 1953.

62. Northern Miner, "Many Companies Active in Sask.," *The Northern Miner*, 1954, Sept. 16, p.2.

63. ITN, *"The Road to Uranium,"* Documentary Film, Independent Television News (ITN), London, U.K., 16 October 1957.

64. British Columbia Geological Survey, "MINFILE Mineral Inventory," MINFILE Record Summary, MINFILE No 082M 021, British Columbia Ministry of Energy and Mines: Victoria, BC, 2013, http://minfile.gov.bc.ca/Summary.aspx?minfilno=082M++021.

65. Piper, L., *The Industrial Transformation of Subarctic Canada*, UBC Press: Vancouver, 2009.

66. Gunnar Mines Ltd., *The Gunnar Story*, Gunnar Mines Ltd., Toronto, Sept., 1957.

67. Delaney, G., "Uranium in Saskatchewan, Canada," Proc. South Australian Resources and Energy Investment Conference (SAREIC 2009), Unlocking South Australia's Mineral Wealth, 6 May 2009, http://www.pir.sa.gov.au/__data/assets/pdf_file/0006/104559/Gary_Delaney.pdf.

68. Schramm, L.L., *Gunnar Uranium Mine: Canada's Cold War Ghost Town*, Saskatchewan Research Council, Saskatoon, 2016.

69. Gunnar Mines, "25th Annual Report. For the Year 1958," Gunnar Mines Ltd., Toronto, 23 March, 1959.

70. SRC, *12th Annual Report of the Saskatchewan Research Council 1958*, Saskatchewan Research Council, Regina, 1959.

71. Cipriani, A.J., "Radiation Hazards in Uranium Mines," *Can. Mining J.*, **1955**, *76(9)*, 79-80.

72. Schramm, L.L., *Research and Development on the Prairies. A History of the Saskatchewan Research Council*, Saskatchewan Research Council, Saskatoon, 2016.

73. Muldoon, J.A., "Policy Networks, Policy Change and Causal Factors, A Uranium Mining Case Study," Ph.D. Thesis, University of Regina, Regina, SK, March 31, 2015.

74. World Nuclear Association, *"Uranium in Canada,"* World Nuclear Association: London, U.K., January, 2016, http://www.world-nuclear.org/info/country-profiles/countries-a-f/canada--uranium.

75. Natural Resources Canada, "About Uranium," Natural Resources Canada, Ottawa, January, 2016, http://www.nrcan.gc.ca/energy/uranium-nuclear/7695.

76. Robinson, S. C., "Mineralogy of the Goldfields District, Saskatchewan, Interim Account" Paper 50-16, Geological Survey of Canada, Ottawa, 1950.

77. Christie, A.M., "Gold Fields and Martin Lake Map-Areas, Saskatchewan," Paper 49-17, Geological Survey of Canada, Ottawa, 1949.

78. Lang, A.H., "Canadian Deposits of Uranium and Thorium, Interim Account" Economic Geology Series No. 16, 1st Ed., Geological Survey of Canada, Ottawa, 1952.

79. Geological Survey, "Goldfields – Martin Lake , Saskatchewan" Map 1015A, Geological Survey of Canada, Ottawa, 1952.

80. Geological Survey, "Goldfields, Saskatchewan" Map 436G, Geological Survey of Canada, Ottawa, 1956.

81. Clifton Associates Ltd., "An Assessment of Abandoned Mines in Northern Saskatchewan (Year 2)," Report for Saskatchewan Environment, Regina, May, 2002.

82. Christie, A.M., "Goldfields-Martin Lake Map-Area, Saskatchewan," Memoir 269, Geological Survey of Canada, Ottawa, 1953.

83. Lang, A.H.; Griffith, J.W.; Steacy, H.R., "Canadian Deposits of Uranium and Thorium," Economic Geology Series No. 16, 2nd Ed., Geological Survey of Canada, Ottawa, 1962.

84. Tyrrell, J.B., "Report on the Country between Athabasca Lake and Churchill River," Report, Geological Survey of Canada, Ottawa, 1896.

85. Jensen, K.A., "Technical Report on the Uranium City Properties for Uranium City Resources Inc. in the Uranium City Area NTS Map Sheets 74N-07, 74N-08, 74N-09 and 74N-10 and the Tazin Lake Area NTS Map Sheets 74N-14 and 74N-15 Northern Mining District Saskatchewan, Canada," Red Rock Energy Inc., Calgary, 2005, http://www.redrockenergy.ca/Technical_Report_-_Uranium_City_Properties.pdf.

86. Allan, J.A.; Cameron, A.E., "An Occurrence of Iron on Lake Athabasca," Report No. 7, Scientific and Industrial Research Council of Alberta, Edmonton, 1923.

87. Lang, A.H., "History of Uranium Discoveries, Lake Athabasca," *Can. Min. J.*, **1953**, *74(6)*, 69-76.

88. Robinson, S.C., Mineralogy of Uranium Deposits, Goldfields, Saskatchewan," Bulletin 31, Geological Survey of Canada, Ottawa, 1955.

89. Harper, C.T., "Northern Iron Ore Exploration," in "Summary of Investigations 1976," Report, Saskatchewan Geological Survey, Regina, 1976, pp. 104-114.

90. Horan, J.W., *On the Side of the Law: Biography of J.D. Nicholson*, Institute of Applied Art Ltd., Edmonton, AB, 1944.

91. SICC, "Mining Impact in Saskatchewan (Timeline)," Saskatchewan Indigenous Cultural Centre, Saskatoon, 2016, http://www.sicc.sk.ca/archive/heritage/ethnography/dene/resources/mining.html.

92. Eldorado, 1948 Annual Report, Eldorado Mining and Refining Ltd., Ottawa, 1949.

93. Wright, J., "Saskatchewan's North," *Can. Geog. J.*, **1952**, *45(1)*, 14-33.

94. Northern Miner, "Record Tonnage Moves Down North to Uranium Mines," *The Northern Miner*, 1956, Oct. 4, p.17,20.

95. Gunnar Mining Ltd., "The Gunnar Story," *Can. Mining J.*, **1963**, *7*, 53-119.

96. Quiring, D.M., *CCF Colonialism in Northern Saskatchewan: Battling Parish Priests, Bootleggers, and Fur Sharks*, UBC Press, Vancouver, 2004.

97. Eldorado, 1949 Annual Report, Eldorado Mining and Refining Ltd., Ottawa, 1950.

98. Eldorado, "Annual Report 1960," Eldorado Mining and Refining Ltd., Ottawa, 1961.

99. Free Press, "High Grade Pitchblende Find Reported at Lake Athabasca," *Free Press* (Winnipeg), **1935**, *Aug. 10*.

100. Globe and Mail, "Nicholson Mines' Holdings in N.W.T. Reviewed," *The Globe and Mail*, **1941**, *May 15*.

101. Globe and Mail, "Nicholson Plans Northwest Work," *The Globe and Mail*, **1940**, *May 11*, p. 21.

102. South Peace Historical Society, "J.D. Nicholson's Problems in the Peace," Article, 13-008, South Peace Historical Society, Dawson Creek, B.C., 2017, http://calverley.ca/article/13-008-j-d-nicholsons-problems-in-the-peace/.

103. The resignation of Asst. Supt. Nicholson from the Alberta Provincial Police is noted in several newspaper articles of the time, including: *Edmonton Journal*, 1 March 1918, 15 March 1918, and *Edmonton Bulletin*, 14 March 1918.

104. RCMP, "Obituary," *Royal Canadian Mounted Police Quarterly*, **1945**, *11(1)*, 79.

105. RCMP, "National RCMP Graves Website & Database," Royal Canadian Mounted Police, Ottawa, 2007, www.rcmpgraves.com/.

106. Lin, Z., *Policing the Wild North-West*, University of Calgary Press, Calgary, 2007.

107. Harding, J.W.; Pedersen, O.E.; Schinkel, R., "Eldorado Beaverlodge Operation," *Can. Mining J.*, **1960**, *81(6)*, 75-149.

108. Northern Miner, "Nicholson Works in Territories," *The Northern Miner*, **1944**, Nov. *2*.

109. Northern Miner, "Nicholson Outlines Work for Current Year," *The Northern Miner*, **1946**, June *20*.

110. Globe and Mail, "Nicholson Awaits Additional Assays on Uranium Show," *The Globe and Mail*, **1948**, *Oct 27*, p. 24.

111. Globe and Mail, "W.B. Milner Heads Nicholson Mines," *The Globe and Mail*, **1948**, *Nov. 1*, p. 23.

112. News, "Encouraging Report from Nicholson," *News of the North (Yellowknife)*, **1948**, *Oct. 15*.

113. Northern Miner, "Nicholson Assays Show High Gold," *The Northern Miner*, **1948**, *Nov. 18*, pp. 1,16.

114. Northern Miner, "Nicholson Reports Uranium in Adit," *The Northern Miner*, **1948**, *Nov. 13*, pp. 1,16.

115. Mamen, C., "Uranium Mining Methods," *Can. Mining J.*, **1956**, *77(6)*, 89-107, 156-157.

116. Globe and Mail, "Nicholson Drills Show Uranium," *The Globe and Mail*, **1949**, *Apr. 27*, p. 24.

117. Globe and Mail, "Nicholson Mines has Good Samples," *The Globe and Mail*, **1950**, *Jan. 24*, p. 20.

118. Globe and Mail, "Nicholson Mine Considers Mill," *The Globe and Mail*, **1950**, *Aug. 17*, p. 18.

119. Globe and Mail, "Nicholson Adds Depth to Zone," *The Globe and Mail*, **1950**, *Aug. 5*, p. 14.

120. MacDonald, B.C.; Kermeen, J.S., "The Geology of Beaverlodge," *Can. Mining J.*, **1956**, *77(6)*, 80-83, 156.

121. Eldorado, 1954 Annual Report, Eldorado Mining and Refining Ltd., Ottawa, 1955.

122. Scott, J., "Athabasca Uranium: Gunnar Kindles Blaze As Eldorado Ace Mine Nears Producing Stage," *The Globe and Mail*, **1953**, *Feb. 9*, p. 25.

123. Globe and Mail, "Five Producers in 1954: More Uranium Output Seen by Saskatchewan," *The Globe and Mail*, **1953**, *Dec. 24*, p. 15.

124. New York Times, "Uranium boom on in Saskatchewan," *New York Times*, **1953**, *May 28*, p. 41.

125. Scott, J., "Lake Cinch Is Newest Uranium Producer," *The Globe and Mail*, **1957**, *Sept. 21*, p. 41.

126. MacDonald, J.R., "The Goldfields Uranium Area," *Western Miner*, **1953**, *26(4)*, 38-45.

127. Argus, "Canada Looks to Active Uranium Year," *The Argus (Melbourne, Victoria)*, **1954**, *Jan. 9*, p. 17.

128. Courier-Mail, "Canada has a 'Rum Jungle'," *The Courier-Mail (Brisbane, Queensland)*, **1954**, *Aug. 3*, p. 2.

129. Life, "History's Greatest Metal Hunt," *Life Magazine*, **1955**, *38(21)*, May 23, p. 25-35.

130. Life, "The Uranium Rush is on in Athabaska," *Life Magazine*, **1954**, *33(7)*, Aug. 18, p. 15-19.

131. Globe and Mail, "Nicholson Mines Leases 54 Claims from Eldorado," *The Globe and Mail*, **1955**, *Sept. 3*, p. 20.

132. Eldorado, 1955 Annual Report, Eldorado Mining and Refining Ltd., Ottawa, 1956.

133. Eldorado, 1956 Annual Report, Eldorado Mining and Refining Ltd., Ottawa, 1957.

134. Robinson, S.C.; Evans, H.T.; Schaller, W.T.; Fahey, J.J., "Nolanite, A New Iron-Vanadium Mineral from Beaverlodge, Saskatchewan," *Am. Miner.*, **1957**, *42*, 619-628.

135. Kennedy, E.G., *Oh Well, What the Hell!*, an autobiography, produced by BookBound Publishing, Macksville, New South Wales, 2002, but apparently never formally published.

136. Globe and Mail, "Control of Nicholson Acquired by Lamaque," *The Globe and Mail*, **1964**, *Jul. 28*, p. B2.

137. Hassard, F.R., "Mineral Lease 5131. Nicholson Bay, Lake Athabasca, Athabasca Mining District, Saskatchewan" Trigg, Wollett & Assoc. Ltd. report for Imperial Oil Ltd., May, 1975.

138. McGill, "Canadian Corporate Reports," Digital Archive, McGill University, Montreal, 2005, http://digital.library.mcgill.ca/hrcorpreports/search/detail.php?compan y=Chancellor%20Energy%20Resources%20Inc.&ID=734.

139. Geo-Help, "Canadian Oil and Gas Companies," Geo-Help Inc., 2011, http://geohelp.net/membermwebsites.html.

140. Ellsworth, H.V., "Lead-Uranium Ratios of Two Saskatchewan Pitchblendes," in "Report of the Committee on the Measurement of Geologic Time," Lane, A.C.; Marble, J.P., National Research Council, Washington, December, 1943, pp. 37-38.

141. Dawson, K.R., "Petrology and Red Coloration of Wall-Rocks, Radioactive Deposits, Goldfields Region, Saskatchewan," Bulletin 33, Geological Survey of Canada, Ottawa, 1956.

142. Sibbald, T.I.I., "Uranium Metallogenic Studies: Nicholson Bay Area," in Summary of Investigations, Misc. Rept. 81-4, Saskatchewan Geological Survey, Regina, 1982, pp. 43-45.

143. Dawson, K.R., "A Petrographic Description of the Wall-Rocks and Alteration Products Associated with Pitchblende-Bearing Veins in the Goldfields Region, Saskatchewan, Preliminary Account," Paper 51-24, Geological Survey of Canada, Ottawa, 1951.

144. Sibbald, T.I.I., "Uranium Metallogenic Studies: Nicholson Bay Area," in Summary of Investigations 1982, Miscellaneous Report 82-4, Saskatchewan Geological Survey, Regina, 1982, pp. 43-45.

145. Sibbald, T.I.I., "Nicholson Bay Uranium-Gold-Platinum Group Element Deposit Studies," in Summary of Investigations 1988, Miscellaneous Report 88-4, Saskatchewan Geological Survey, Regina, 1988, pp. 77-81.

146. Joliffe, A.W., "The Gunnar 'A' orebody," *CIM Trans.*, *1956*, 59, 181-185.

147. Ogilvie-Evans, P., Personal communication, SRC internal email, 1 Feb. 2017.

148. Globe and Mail, "Nicholson Resumes Shipments," *The Globe and Mail*, **1956**, *Jan. 5*, p. 28.

149. Crawford, J.E., "Uranium," in *Minerals Yearbook, 1955*, U.S. Bureau of Mines, Washington, 1958, pp. 1213-1242.

150. Crawford, J.E.; Paone, J., "Uranium," in Minerals Yearbook, 1956, U.S. Bureau of Mines, Washington, 1958, pp. 1245-1290.

151. Crawford, J.E., "Uranium and Radium," in *Minerals Yearbook, 1954*, U.S. Bureau of Mines, Washington, 1958, pp. 1241-1296.

152. Saskatchewan Research Council, KHS Environmental Management Group, and CanNorth Environmental Services, "Gunnar Site Characterization and Remedial Options Review," SRC Publication No. 11882-1C04, Saskatoon, January, 2005.

153. Gunnar Mining Ltd., "The Gunnar Story," *Can. Mining J.*, **1963**, *7*, 53-119.

154. Beckett, T., "Goldfields: The First Boom Town," Blog post, Uranium City History, 2016, http://uraniumcity-history.com/places/goldfields/.

155. UER.CA, "Uranium City," Urban Exploration Resource, UER.CA, 2014, http://www.uer.ca/locations/show.asp?locid=23608.

156. Horton, A.G., "Report on Activity at Nicholson," Geological Activity report, 26 June 1950.

157. Grant, N.A., "General Mining Conditions at Eldorado Beaverlodge," *Trans. Can. Inst. Min. Met.*, **1953**, *56*, 260-263.

158. Butler, R.D., "Carbonate Leaching of Uranium Ores. A Review," Paper III, Proc. AAEC Symp. on Uranium Processing, Australian Atomic Energy Commission, Lucas Heights, Australia, 20-21 July, 1972 pp. III-1 to III-3.

159. Thunaes, A., "Uranium Recovery Plants," *Can. Mining J.*, **1956**, *77(6)*, 123-126,159.

160. Hannay, R.L., "Milling at Beaverlodge," *Can. Mining J.*, **1956**, *77(6)*, 135-140, 161.

161. Hornsey, D.; Lee, R.G.H., "Hydrometallurgical Leaching Process and Apparatus," U.S. Patent 5,250,273, 5 Oct. 1993.

162. Gunnar Mines, "*Gunnar Progress*," Documentary Film, Gunnar Mines Ltd., Toronto, *circa*. 1958.

163. Burger, J.C.; Jardine, J.McN., "Canadian Refining Practice in the Production of Uranium Trioxide by Solvent Extraction with Tributyl Phosphate," Proc. 2nd United Nations Internat. Conf. on the Peaceful Uses of Atomic Energy, Geneva, June, 1958, 17 pp.

164. World Nuclear Association, "How Uranium Ore is Made into Nuclear Fuel," World Nuclear Association, London, UK, 2016, http://www.world-nuclear.org/nuclear-basics/how-is-uranium-ore-made-into-nuclear-fuel.aspx.

165. Pitkanen, L.L., "A Hot Commodity: Uranium and Containment in the Nuclear State," Ph.D. Thesis, Department of Geography, University of Toronto, Toronto, 2014.

166. Globe and Mail, "New Uranium Refining Plan Being Installed by Eldorado," *The Globe and Mail*, **1954**, *Sept. 8*, p. 24.

167. Eldorado, "Annual Report 1958," Eldorado Mining and Refining Ltd., Ottawa, 1959.
168. Page, R.D.; Lane, A.D., "The Performance of Zirconium Alloy Clad UO$_2$ Fuel for Canadian Pressurized and Boiling Water Power Reactors," *Proc. Joint ANS-CNA Conference*, Toronto, June 10-12, 1968, Canadian Nuclear Association, Ottawa, Paper AECL-3068.
169. Saskatchewan Environment, "Northern Operations –Abandoned," Field Note Book #017, 1989-1993, Saskatchewan Environment, Regina, 1993.
170. Waggitt, P., "Uranium Mining Legacy Sites and Remediation - A Global Perspective," Presented at: IAEA Conference, Namibia, October, 2007, International Atomic Energy Agency, http://www.iaea.org/OurWork/ST/NE/NEFW/documents/RawMaterials/CD_TM_Swakopmund%20200710/13%20Waggit4.PDF.
171. IAEA, "Advancing Decommissioning and Environmental Remediation Programmes," IAEA Nuclear Energy Series Report No. NW-T-1.10, International Atomic Energy Agency, Vienna, 2016.
172. CNSC, "Uranium Mines and Mills Waste," Canadian Nuclear Safety Commission, Ottawa, 2014, http://www.nuclearsafety.gc.ca/eng/waste/uranium-mines-and-millswaste/index.cfm.
173. Larmour, A., "Elliot Lake Hailed as Reclamation Success Story," Sudbury Mining Solutions J., 2010, Sept. 1, http://www.sudburyminingsolutions.com/elliot-lake-hailed-as-reclamation-success-story.html.
174. AANDC, "Port Radium Mine (Remediation Complete)," Aboriginal Affairs and Northern Development Canada, Ottawa, 2012, http://www.aadnc-aandc.gc.ca/eng/1332423218253/1332441057035.
175. Bihun, G., "Lorado Mine Historical File," Saskatchewan Environment, Regina, May, 2015.
176. Golder Associates Ltd., "Technical Information Document for the Inactive Lorado Uranium Tailings Site," Report for EnCana Corp., April, 2008.
177. Wilson, I.; Allen, D.E.; Schramm, L.L.; Muldoon, J., "Lorado Uranium Mine Environmental Remediation – Northern Saskatchewan," *Proc. IAEA Internat. Conf. on Advancing the Global Implementation of Decommissioning and Environmental Remediation Programmes*, Madrid, Spain, May 23-27, 2016, paper IAEA-CN-238-30P.
178. SRC, "Project CLEANS (Cleanup of Abandoned Northern Sites)," Saskatchewan Research Council, Saskatoon, 2014, http://www.src.sk.ca/about/featured-projects/pages/project-cleans.aspx.
179. Castrilli, J.F., "Wanted: A Legal Regime to Clean Up Orphaned /Abandoned Mines in Canada," *J. Sust. Devel. Law Policy* **2010**, *6(2)*,109-141.

180. Joint Research Centre, "Advancing Implementation of Nuclear Decommissioning and Environmental Remediation Programmes," Policy Support Document EUR 27902, European Commission, Brussels, 2016.

181. Brown, L.D., "Proposed Decommissioning of the Gunnar and Lorado Uranium Mine Sites," Report for Saskatchewan Environment, BB Health Physics Services, Regina, SK, 1993.

182. Stenson, R.; Howard, D., "Regulatory Oversight of the Legacy Gunner Uranium Mine and Mill Site in Northern Saskatchewan, Canada – 13434," *Proc., Waste Management Conference (WM2013)*, WM Symposia, Inc., Tempe, AZ, 2013, 13 pp.

183. Peach, I.; Hovdebo, D., "Righting Past Wrongs: The Case for a Federal Role in Decommissioning and Reclaiming Abandoned Uranium Mines in Northern Saskatchewan," Public Policy Paper 21, Saskatchewan Institute of Public Policy, University of Regina, December, 2003.

184. Saskatchewan, "Canada's New Government and Province of Saskatchewan Launch First Phase of Cleanup of Legacy Uranium Mines," News Release, Government of Saskatchewan, Regina, April 2, 2007.

185. Editorial Board, "U.C. Meeting on Abandoned Mines," Supplement, *Opportunity North*, **2008**, *Spring*, 2-3.

186. Editorial Board, "Cleanup Process Begins at Gunnar," *Opportunity North*, **2009**, *Summer*, 22.

187. Editorial Board, "Project CLEANS Team Gears up for a New Work Season," *Opportunity North*, **2013**, *Spring*, 29 (see also p. 20).

188. Petelina, E., Sanscartier, D.; MacWilliam, S.; Ridsdale, R., "Environmental, Social, and Economic Benefits of Biochar Application for Land Reclamation Purposes," *Proc. 38th Ann. B.C. Mine Reclamation Symposium*, 2014, 13 pp.

189. Saskatchewan, "Northern Saskatchewan Environmental Quality Committee," Government of Saskatchewan, Regina, 2016, https://www.saskatchewan.ca/residents/first-nations-citizens/saskatchewan-first-nations-metis-and-northern-initiatives/northern-saskatchewan-environmental-quality-committee.

190. Saskatchewan Research Council, "Gunnar Site Remediation Project: Environmental Impact Statement," SRC 12194-320-1L13, February, 2013.

191. Provost, K., "Abandoned Mine Being Cleaned-Up in Saskatchewan," CJLR-FM News, La Ronge, SK, broadcast 25 October 2010.

192. Muldoon, J.; Schramm, L.L., "Gunnar uranium mine remediation project. Northern Saskatchewan," *Proc. 33rd Arctic and Marine Oilspill Program (AMOP) Technical Seminar on Environmental Contamination and Response*, Halifax, N.S., June 7-9, pp. 383-403, 2010.

193. Muldoon, J.; Schramm, L.L., "Gunnar Uranium Mine Environmental Remediation – Northern Saskatchewan," Paper ICEM2009-16102, *Proc. 12th Internat. Conf. Environmental Remediation and Radioactive Waste Management - ICEM'09/DECOM'09*, Liverpool, U.K., October 11-15, 2009.

194. Redmann, R.E.; Frankling, F.T., "Revegetation of Abandoned Uranium Mill Tailings Near Uranium City, Saskatchewan. Plant Species Selection," Report for Saskatchewan Environment, Regina, March, 1982.

195. Mawhiney, A-M.; Pitblado, J. (Eds.), *Boom Town Blues: Elliot Lake, Collapse and Revival in a Single Industry Community*, Dundurn Press, Toronto, 1999.

196. MacPherson, A. "Gunnar Cleanup to Exceed $250M, 10 Times Estimate," *Saskatoon StarPhoenix*, October 17, 2015, Last Updated: February 18, 2016.

197. MacPherson, A. "Overbudget Gunnar Cleanup Federal Responsibility, Sask. Politicians Say," *Saskatoon StarPhoenix*, February 24, 2016.

198. AAEC, "Rum Jungle Project," Booklet, Australian Atomic Energy Commission, Lucas Heights, Australia, 1963.

199. World Nuclear Association, "Former Australian Uranium Mines," World Nuclear Association, London, U.K., 2014, http://www.world-nuclear.org/info/Country-Profiles/Countries-A-F/Appendices/Australia-s-former-uranium-mines/.

Also of interest …

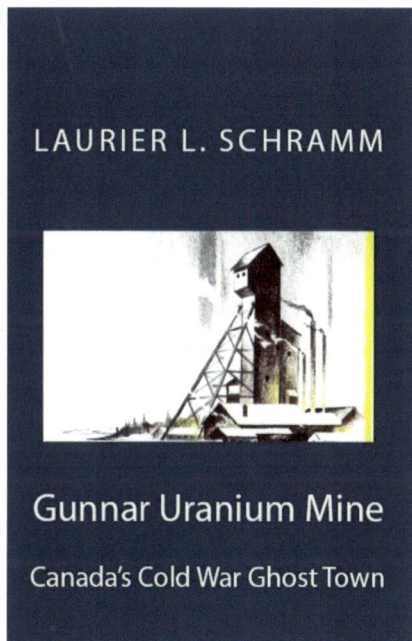

LAURIER L. SCHRAMM

Gunnar Uranium Mine

Canada's Cold War Ghost Town

The Gunnar mine, mill, and town-site were built in a remote location in northern Saskatchewan, on the shore of Lake Athabasca. Like most mining communities the town boomed, first with construction workers and miners, and later with families. When the Gunnar mill construction was completed in the fall of 1955 it doubled Canada's uranium production capacity. By 1956 the Gunnar mine was the largest uranium producer in the world. The Gunnar town-site was built to serve the mine and mill and at one time had a population of about 850 people. By 1964 it was a ghost town. The Gunnar mine produced over 5 million tonnes of uranium ore, nearly 4.4 million tonnes of mine tailings, and an estimated 2,710,700 m^3 of waste rock. Following closure in 1964, the Gunnar site was abandoned with little remediation and no reclamation being done. It has been referred-to as "the second greatest environmental disaster area in Canada." Forty years would pass before the governments of Saskatchewan and Canada reached an agreement to fund the remediation (clean-up) of the Gunnar site, and contracted the management of the project to the Saskatchewan Research Council (SRC). At the time of writing this book the clean-up was well underway, with several years of clean-up activity remaining, and a further expected 10-15 years of monitoring activity before the site is expected to be released into a long-term management and monitoring program.

Print ISBN: 978-0-9958081-2-6   **ePub ISBN: 978-0-9958081-0-2**

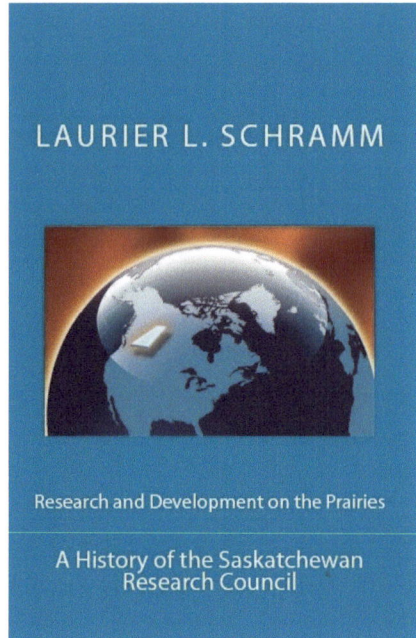

Early in the 20th century, the advent of industrial research councils brought organized research, development, and technological innovation to North America. Such organizations, now usually referred to as research and technology organizations (RTOs), focused on research and development aimed at helping industry develop and advance, and they were critical to the evolution of the modern approach to technological innovation. One of Canada's first RTOs to be established was the Saskatchewan Research Council (SRC), and it has become one of the most enduring. This book traces the evolution of SRC from its first efforts in the 1930s through several distinct eras.

Print ISBN: 978-0-9958081-3-3   **ePub ISBN: 978-0-9958081-1-9**

www.ingramcontent.com/pod-product-compliance
Lightning Source LLC
Chambersburg PA
CBHW042147220326
41599CB00003BB/15